P R A I S E _____ *'eople*

"Talk about down-and-dirty. Or rather, down-and-clean! Here's the actual useful detail on how to do the stuff that really needs doing. Read it and get to work!"
—BILL MCKIBBEN, author of *Eaarth: Making a Life on a Tough New Planet*; founder of 350.org

"Greg Pahl's *Power from the People* is an inspirational guide to the burgeoning community-power movement. His case studies of people who are making a difference are often tales of endurance and survival, but also powerful testaments to the human spirit. Bravo to Pahl and *Power from the People* for explaining how feed-in tariffs have produced a community-power revolution in Europe and how they can do the same here in North America."
—PAUL GIPE, author of *Wind Power*; renewable-energy advocate and industry analyst

"Energy is at the heart of our 21st-century economic-ecological crisis, but most writing on the subject is suffused either with immobilizing anticipation of doom or giddy wishful thinking. Here at last is a genuinely helpful energy book, one that's realistic and practical. If you want to actually do something about our energy future, here is where to start."
—RICHARD HEINBERG, senior fellow, Post Carbon Institute; author of *The End of Growth*

"Greg Pahl's superb guide to community energy and how to unlock its potential is essential reading for anyone interested in the economic future of the place where they live. As a Community Resilience Guide it is just that—a powerful guide showing how enhancing your community's resilience is a key form of economic development. And there is nowhere better to start than with rethinking our relationship to energy. You will find yourself waking up at 3 AM to scribble down the ideas and actions that this book has inspired. Make sure you keep a pen and paper by the bed."

—ROB HOPKINS, author of *The Transition Companion*

POWER FROM THE PEOPLE

HOW TO ORGANIZE, FINANCE, AND

LAUNCH LOCAL ENERGY PROJECTS

GREG PAHL

Foreword by VAN JONES

A COMMUNITY RESILIENCE GUIDE

POST CARBON INSTITUTE
SANTA ROSA, CALIFORNIA

CHELSEA GREEN PUBLISHING
WHITE RIVER JUNCTION, VERMONT

green press INITIATIVE

Copyright © 2012 Post Carbon Institute.
All Rights Reserved.
No part of this book may be transmitted or reproduced in any form by any means without permission in writing from the publisher.

Editor: Joni Praded
Developmental Editor: Daniel Lerch
Copy Editor: Laura Jorstad
Proofreader: Eric Raetz
Indexer: Shana Milkie
Designer: Melissa Jacobson

Printed in the United States of America
First printing July 2012
10 9 8 7 6 5 4 3 2 1   12 13 14 15 16

Chelsea Green Publishing is committed to preserving ancient forests and natural resources. We elected to print this title on 30-percent postconsumer recycled paper, processed chlorine-free. As a result, for this printing, we have saved:

22 Trees (40' tall and 6-8" diameter)
9 Million BTUs of Total Energy
2,209 Pounds of Greenhouse Gases
9,960 Gallons of Wastewater
632 Pounds of Solid Waste

Chelsea Green Publishing made this paper choice because we and our printer, Thomson-Shore, Inc., are members of the Green Press Initiative, a nonprofit program dedicated to supporting authors, publishers, and suppliers in their efforts to reduce their use of fiber obtained from endangered forests. For more information, visit: www.greenpressinitiative.org.

Environmental impact estimates were made using the Environmental Defense Paper Calculator. For more information visit: www.papercalculator.org.

Our Commitment to Green Publishing

Chelsea Green sees publishing as a tool for cultural change and ecological stewardship. We strive to align our book manufacturing practices with our editorial mission and to reduce the impact of our business enterprise in the environment. We print our books and catalogs on chlorine-free recycled paper, using vegetable-based inks whenever possible. This book may cost slightly more because it was printed on paper that contains recycled fiber, and we hope you'll agree that it's worth it. Chelsea Green is a member of the Green Press Initiative (www.greenpressinitiative.org), a nonprofit coalition of publishers, manufacturers, and authors working to protect the world's endangered forests and conserve natural resources. *Power from the People* was printed on FSC®-certified paper supplied by Thomson-Shore that contains at least 30% postconsumer recycled fiber.

Library of Congress Cataloging-in-Publication Data
Pahl, Greg.
  Power from the people : how to organize, finance, and launch local energy projects / Greg Pahl ; foreword by Van Jones.
       p. cm. — (A community resilience guide)
  Includes bibliographical references and index.
  ISBN 978-1-60358-409-8 (pbk.) — ISBN 978-1-60358-410-4 (ebook)
1. Small power production facilities. 2. Community development. I. Title.

  TK1006.P34 2012
  307.1'4—dc23
                              2012016893

Chelsea Green Publishing
85 North Main Street, Suite 120
White River Junction, VT 05001
(802) 295-6300
www.chelseagreen.com

The fate of people on Earth depends
on whether we can employ efficient
and renewable energies. We need
to lay big plans for small technologies.

—DAVID FREEMAN, former head of the New York Power Authority,
Tennessee Valley Authority, Sacramento Municipal Utility District,
and Lower Colorado River Authority, speaking at the
World Renewable Energy Congress in June 1996

Introduction to the Community Resilience Guides

In the twenty-first century, we face a set of interconnected economic, energy, and environmental crises that require all the courage, creativity, and cooperation we can muster. These crises are forcing us to fundamentally rethink some of our most basic assumptions, like where our food and energy come from, and where we invest our savings.

While national and international leadership are key to navigating the bumpy road ahead, that leadership thus far is sadly wanting. And, in any case, many of the best responses to these challenges are inherently local.

Thankfully, a small but growing movement of engaged citizens, community groups, businesses, and local elected officials are leading the way. These early actors have worked to reduce consumption, produce local food and energy, invest in local economies, and preserve local ecosystems. While diverse, the essence of these efforts is the same: a recognition that the world is changing and the old way of doing things no longer works.

Post Carbon Institute has partnered with Chelsea Green Publishing to publish this series of Community Resilience Guides to detail some of the most inspiring and replicable of these efforts. Why community resilience? *Community*, because we believe that the most effective ways to work for the future we want are grounded in *local* relationships—with our families and neighbors, with the ecological resources that sustain us, and with the public institutions through which we govern ourselves. *Resilience*, because the complex economic, energy, and environmental challenges we face require not solutions that make problems go away but *responses* that recognize our vulnerabilities, build our capacities, and adapt to unpredictable changes.

These are frightening, challenging times. But they are also full of opportunity. We hope these guides inspire you and help you build resilience in your own community.

Asher Miller
Executive Director
Post Carbon Institute

CONTENTS

by Van Jones

This book rests an optimistic message on a pessimistic premise.

The sobering underlying thesis is that human civilization is already in big trouble—both ecologically and economically. And things are set to get much worse. The hopeful underlying message is that we still have the capacity to pull good outcomes from even the most frightening scenarios.

The paradox is this: Only by recognizing how much worse things can get can we muster the energy and creativity to win a better future. In that regard, the book you hold in your hands is not just an action guide; it is a survival guide.

The Bad News Is Very Bad

At this late date, there is no point in mincing words about the impending series of calamities. The global production of oil will soon peak, ending forever the era of cheap crude. The resulting price spikes and fuel shortages could throw all of industrial society into an ugly death spiral. Worse still: We have seen only the earliest examples of the kind of biblical disasters—the super-storms, wildfires, floods, and droughts—that climate experts predict are in the pipeline, even if we cease all carbon emissions immediately.

The polar ice caps haven't melted yet; if they do, they will send temperatures and sea levels soaring, forcing us to redraw every coastal map in the world. Even under the friendliest scenarios, we will likely see food systems disrupted, life-sustaining fuels priced beyond reach for many, and our health challenged as tropical super-bugs invade formerly temperate climes. On a hotter planet, we could face the choice between water rationing and water riots. As stressful as the present moment is, worse times are possible—and even likely.

At the same time, the majority of the world's people now live in cities. And though cities cover only 2 percent of Earth's surface, they already consume 75 percent of the planet's natural resources. As more people continue crowding into cities, that figure will climb even higher, which means urban areas have become the main driver in the ecological crisis. Many cities are sinkholes of human suffering, especially for a marginalized population of low-income earners and people of color. And in the United States, the word *urban* has become synonymous with the word *problem*. Many urban neighborhoods are plagued by economic desperation, violence, pollution, and crumbling infrastructure.

Climate change and the economic and equity crises of our communities may appear to have little in common, but they share a key determining factor—namely, our near-complete dependence on coal, oil, and natural gas. The carbon dioxide produced by driving our vehicles, heating (and cooling) our homes, and lighting our cities with fossil fuels is the main culprit behind climate change. Meanwhile, that same dependence on fossil fuels sucks billions of dollars every year out of communities across America, with the poorest households often hit hardest.

But what if we found ways to power our homes, businesses, factories, and vehicles that *didn't* warm the planet, that kept local dollars circulating in local economies, and that even created local jobs? What if we spread those climate-friendly, local-economy-boosting, job-creating ideas to every city and town across the country?

Hope for the Best, Prepare for the Worst

It is too late for us to avert all of the negative consequences of 150 years of ecological folly and resource wastefulness. Our challenge is to begin implementing real changes, rapidly and from the bottom up. Certain bills are coming due, and certain chickens are coming home to roost, no matter what we do. But there are steps we can take to cushion the blow.

We must prepare ourselves (and our communities) for the worst possible outcomes. In considering the most pessimistic scenarios, we must talk less about economic growth and more about economic resilience; less about abundance and more about sufficiency; less about sustainability and more about survivability. It may be wise to consciously deploy our forces in a three-pronged, "trident" formation: some of us fixing the system from the inside, some of us pressuring the system from the outside, and some of us exercising the "lifeboat" option, thinking up alternative strategies for survival.

Power from the People is rare, because it gives some guidance on all three around the most important component of that system: energy.

You'll read about courageous local government leaders finding creative ways to invest in local renewable energy; citizen activists pushing for (and winning) smarter regulations for green power; and entire communities taking matters into their own hands to prepare for an energy-scarcer future. Throughout the stories here, from both urban and rural communities, you'll find a common theme all too often missing from the sustainability conversation: *local prosperity*. Local renewable energy is the heart of the new energy economy because it is the most obvious starting point for creating green jobs and generating local wealth. Local renewable energy puts the *power* in *local empowerment*.

By itself, however, even the most advanced local energy initiative can do little about our energy and environmental crises. Local actions must be multiplied to the level of movements . . . and nations.

Can America summon the strength, courage, and resolve to avert disaster and usher in a new age of sustainable prosperity? Both the ideas and the constituencies exist to turn the corner. We need a hard-hat-and-lunch-bucket brand of environmentalism . . . a we-can-fix-it environmentalism . . . a muscular, can-do environmentalism. We need a pro-ecology movement with its sleeves rolled up and its tool belt strapped on. We need a social uplift environmentalism that can fight poverty and pollution at the same time—by creating green-collar jobs for low-income people and displaced workers.

The time has come to birth a positive, creative, and *powerful* environmentalism, one deeply rooted in the lives, values, and needs of millions of ordinary people who work every day (or desperately wish they could).

We need an environmental movement that can put millions of people back to work, giving them the tools and the technologies they need to retrofit, re-engineer, and reboot the nation's energy, water, and waste systems. Green-collar jobs can restore hope and opportunity to America's failing middle-class and low-income families while honoring and healing the Earth. Those new jobs could create a ladder up and out of poverty for jobless urban residents. Under even the most depressing of scenarios, there certainly will be economic opportunities and green-collar jobs—from building dikes and levees and reconstructing devastated structures to installing community-owned wind turbines and operating renewable biofuel factories using regional feedstocks. The United States can fight global warming, energy scarcity, and poverty in the same stroke.

With 4 percent of the world's population, the United States now produces 25 percent of the world's greenhouse gas pollution. It also locks up 25 percent of the world's prisoners in its domestic incarceration industry. Those numbers document the notion that too many U.S. business and political leaders govern as if we have both a disposable planet and disposable people.

As the new green economy springs to life, will we live in eco-equity or eco-apartheid? Will clean and green business flourish only in the rich, white parts of town? Will our kids be left to deal with the toxic wastes of polluting industries, the life-threatening diseases that decimate polluted communities, and the crushing lack of economic opportunity as the old polluting economy goes bust? How we answer these questions will impact the fate of billions of people.

On this crowded planet, we have responsibilities that extend beyond our national borders. Therefore, it is good to be a global citizen. But we must never forget: The very best gift that we can give to the world is a better America. The peoples of the world want and need our country to set a global example for human and environmental rights while being a global partner for peace and progress.

We are entering the tough terrain of an unforgiving new century. But there is a path forward. It is narrow and treacherous, but it leads to the best possible outcome for the largest number of people. And it starts with developing local renewable energy.

VAN JONES
JUNE 2012

Acknowledgments

While I worked on this exciting project, I met and spoke with many dedicated, enthusiastic people who are committed to helping the global community free itself from its dependency on fossil fuels, and who wish to build a better, more sustainable society at the community level.

I would like to acknowledge the many people who were so generous with their time and advice. This book would not have been possible without people like Steve Whitman, board member, Plymouth Area Renewable Energy Initiative, Plymouth, New Hampshire; Steve Klein, general manager, Janne Avatare, executive assistant to the general manager, Neil Neroutsos, corporate communications, Snohomish County PUD, Everett, Washington; Rich Carpenter, treasurer, Acorn Renewable Energy Co-op, Middlebury, Vermont; Gary Nystedt, resource manager, City of Ellensburg, Washington; Joy Hughes, founder and CEO, SolarPanelHosting.com, Westminster, Colorado; Don Schramm, Burlington Cohousing, Burlington, Vermont; Noah Pollock, project manager, Community Energy Exchange, Burlington, Vermont; Terry Meyer, owner, Cascade Community Wind Co. LLC, Bellingham, Washington; Jon Folkedahl, president, Folkedahl Consulting, Inc., Willmar, Minnesota; Suzanne Pude, community energy director, Island Institute, Rockland, Maine; Chip Farrington, general manager, Fox Islands Electric Cooperative, Vinalhaven, Maine; Jeff White, former director, Logan City Light & Power, City of Logan, Utah; Cheryl Johnson, public works administrator, Spearfish, South Dakota; Mark Brandenburger, director of special utilities projects, Hamilton, Ohio; Kent Carson, senior director of communications, American Municipal Power, Columbus, Ohio.

I would also like to thank Jim Bishop, CEO, Harney District Hospital, Burns, Oregon; Geoff Battersby, former mayor and energy project coordinator, Woodchip District Heating System, City of Revelstoke, British Columbia, Canada; Tom Corbin, director of business services, Middlebury College, Middlebury, Vermont; Paul Greene, American Biogas Council, Washington, DC; Kevin Maas, president, Farm Power Northwest, Mount Vernon, Washington; Larry Hare, WRF operations supervisor, Des Moines Metro Wastewater

Reclamation Authority, Des Moines, Iowa; John Welch, recycling manager / project manager, Mike DiMaggio, solid waste manager, Dane County Public Works, Madison, Wisconsin; Mark Torresani, engineer, Cornerstone Environmental Group, LLC, Madison, Wisconsin; Lyle Estill, president and founder, Piedmont Biofuels, Pittsboro, North Carolina; Lynn Benander, CEO, Co-op Power and Northeast Biodiesel, Jonathan T. Walsh, project coordinator, Northeast Biodiesel, Greenfield, Massachusetts; Mike Jerke, general manager, Chippewa Valley Ethanol Company, Benson, Minnesota; Delayne Johnson, general manager, Quad County Corn Processors, Galva, Iowa; Tonya "Toni" Boyd, senior engineer, Geo-Heat Center, Oregon Institute of Technology, Klamath Falls, Oregon; Ann Felton, chairman and chief executive, Aaron McRee, construction manager, Central Oklahoma Habitat for Humanity, Oklahoma City, Oklahoma; Mike Antheil, executive director, Florida Alliance for Renewable Energy, West Palm Beach, Florida; Ed Regan, assistant general manager for strategic planning, Kim Jamerson, senior marketing and communications specialist, Gainesville Regional Utilities, Gainesville, Florida; Elizabeth Marcus and Conrad Willeman, Transition Newburyport, Newburyport, Massachusetts; Niall Robinson, member, mayor's Energy Advisory Committee, Newburyport, Massachusetts; and Susi Vogl, council administrator, Wildpoldsried, Bavaria, Germany.

I also want to thank Jake Fischer, clean energy program manager, The Minnesota Project, for permission to use the City of Willmar, Minnesota, case study from *Lessons & Concepts for Advancing Community Wind*, December 2009, as the basis for the City of Willmar case study in this book.

In addition, many thanks to Daniel Lerch, the publications director, and Asher Miller, the executive director, Post Carbon Institute, for their editorial assistance, numerous suggestions, and other support they provided for this project.

I would like to offer my sincere thanks all the wonderful folks at Chelsea Green Publishing who helped to bring this book to completion.

I also want to thank anyone else I may have forgotten to mention. All of your contributions, both large and small, are greatly appreciated.

Last, but by no means least, I want to thank my wife, Joy, for her help in making suggestions, and generally putting up with me while I was trying to meet deadlines.

JANUARY 2012

The monster tornado that touched down in Joplin, Missouri, on the evening of May 22, 2011 cut a path nearly six miles long and three-quarters of a mile wide through the center of the city. One of the worst-hit locations was St. John's Regional Medical Center, which suffered a direct hit. The staff barely had enough time to move patients into hallways before the storm blew out every window, creating chaos throughout the building. Five patients died as a result of the tornado. Debris from the hospital—medical records, X-rays, insulation, and other items—later fell to the ground up to sixty miles away.

The tornado devastated about a quarter of the city, shredding around two thousand buildings, knocking out electricity and cell phone service for many, and damaging water treatment and sewage plants in its path.

Jeff Lehr, a reporter for *The Joplin Globe*, was upstairs in his home when the storm hit but was able to make his way to a basement closet.

"There was a loud huffing noise, my windows started popping. I had to get downstairs, glass was flying. I opened a closet and pulled myself into it," he told the Associated Press. "Then you could hear everything go. It tore the roof off my house, everybody's house. I came outside and there was nothing left."[1]

Other local residents weren't so lucky. The final death toll from the Joplin tornado was at least 160. More than 750 residents were injured. The National Weather Service later said that the tornado was an EF-5 storm—the highest rating—with winds greater than two hundred miles per hour. It was the deadliest tornado in the United States since 1947.

Joplin was not the only community to suffer tornado damage in 2011. Alabama, Mississippi, Tennessee, Georgia, Virginia, Louisiana, Kentucky, and Massachusetts were also hard hit. Collectively, the storms were the second worst in U.S. history and the worst in terms of sheer numbers of tornadoes in such a short period of time: 226 occurred within a twenty-four-hour period on April 27 and 28. Nationwide, the death toll from all tornado activity in 2011 was 552, and the financial cost was around $25 billion.[2]

These unsettling statistics are just part of a much larger picture of extreme weather that is becoming more frequent—and more costly. Record snowfall and rainfall in 2011 in many parts of the nation resulted in record flooding, from the Mississippi River and its tributaries (including the Missouri River) in the West, Midwest, and South, to Lake Champlain between New York and Vermont in the Northeast, and many other locations in between. In August 2011, Hurricane Irene drenched the eastern seaboard and triggered record flooding in New Jersey, New York, and Vermont, causing more than $7 billion in damage. While some regions were underwater, others suffered from record drought. Texas and parts of Oklahoma and New Mexico were drier than they have ever been—worse than during the Dust Bowl. Massive wildfires in these and other states caused most previous fire records to go up in smoke. In 2011, more than seventy-three thousand fires burned across 8.7 million acres in the United States.[3]

Global Warming

Is the growing number of extreme weather events connected to global warming? Most climate scientists around the world are in agreement that the burning of fossil fuels—coal, oil, and natural gas, along with deforestation—is largely to blame for the rise in carbon dioxide levels in the atmosphere, and that this has raised the average temperature of the Earth's near-surface air around 1.4 degrees Fahrenheit since 1880. They are also largely in agreement that the rate of warming is increasing.[4] According to scientists at the National Oceanic and Atmospheric Administration (NOAA), 2010 tied with 2005 as the warmest year of the global surface temperature on record. This was the thirty-fourth consecutive year with global temperatures above the twentieth-century average.[5] While any one weather event cannot be directly linked to climate change, the overall surge in extreme weather events such as wildfires, heat waves, and strong tropical storms is certainly consistent with scientific predictions of the general pattern of climate destabilization to be expected with global warming.

Unfortunately, despite all the talk about reducing it, more carbon dioxide was spewed into the atmosphere as a result of the burning of fossil fuels in 2010 than at any other time in human history. As if to underscore that fact, the past few years also saw unprecedented flooding in Australia, New Zealand, and Pakistan, to say nothing of dramatically accelerated Arctic and

Antarctic ice melt, nor the massive drought-related crop failures in Russia, France, and Germany. And then there is the Amazon, which has suffered two "hundred-year" droughts in the past five years.

"The term '100-year event' really lost its meaning this year," U.S. Federal Emergency Management Agency director Craig Fugate said in December 2010, referring to record flooding in the Mississippi and elsewhere in recent years.[6] Worst of all, this is just a preview of what's to come. It all adds up to a troubling picture of global climate destabilization on a massive scale, and big trouble for the global economy—and everyone who depends on it.

The Age of Extreme Energy

If that were all we had to worry about, it would be more than enough. But then there is the peak oil problem. The peaking of global oil production is a huge threat because our modern global economy is almost totally dependent on relatively cheap oil and petroleum-based products.

Peak oil occurs when the all-time maximum rate of petroleum extraction is reached. There is a growing consensus that this has already occurred and that we are currently bumping along the top of the so-called Hubbert curve[7] of global oil production, about to begin a slow but inevitable decline. Even the traditionally over-optimistic International Energy Agency says that production of "conventional" oil peaked in 2006.[8]

The same peaking problem generally applies to natural gas and coal as well. The schedule of their eventual peaking and decline may not be quite as tight as for oil, but it is just as inevitable—and equally serious. We are entering the era of extreme energy.

The combined impact of global climate destabilization and extreme energy is setting the stage for a global catastrophe. All of this could eventually bring the global economy to its knees. Higher and more volatile fossil fuel prices will increase the cost or create shortages of the many products that we have come to rely on in our daily lives. The end of cheap oil and natural gas will undoubtedly cause serious disruptions in the production of food, causing prices to skyrocket—and we can expect even higher food prices and tighter food supplies as climate destabilization causes further crop failures. Prices for most other goods will also increase dramatically as rising demand for fossil fuels continues to bump up against an irreversible decline in supply.

Doing Nothing

Unfortunately, our political leaders, especially in Washington, DC, still don't get it. Republicans (and many Democrats) seem to think that burning more coal, oil, and natural gas, along with increased reliance on nuclear power, is somehow going to solve our energy dilemma. The Great Recession and its painfully slow recovery have caused an already deeply divided and partisan Congress to sink into poisonous and mindless theatrics where virtually nothing useful is accomplished. Don't expect anything useful from inside the Beltway anytime soon.

It should be clear from all of this that we face some enormous challenges that are only going to get worse in the years ahead. That's the new, twenty-first-century reality. We need to accept that there are limits to growth, and that we have already passed those limits. We need to understand that the high-consumption global economy is unsustainable, and that the interconnected international monetary system it depends upon to function is extremely fragile—to say nothing about our endangered global ecosystems, which are similarly being degraded "at a rate unprecedented in human history" according to the World Wildlife Fund. We face challenges of biblical proportions, and it's increasingly clear that the federal government is incapable of taking any sort of bold leadership role in planning for how to deal with these looming challenges. Consequently, we need to take action now at the local level to begin to provide for our basic needs from local resources. But how to do that?

Building Resilience

The main strategy can be summed up in a single word: resilience. In this context, resilience means the ability of a person or community to adapt to changing and uncertain circumstances. We need to build greater resilience at the local level, both individually and collectively, so that we will be able to respond to the wide range of energy, food, and economic problems we will have to deal with—simultaneously. This will require that we be nimble, quick to respond and adapt to rapidly changing circumstances, and courageous and cooperative in our responses. The communities that thrive in the twenty-first century will be the ones that best learn how to adapt and work together; the phrase *It takes a village to raise a child* will take on a whole new meaning. We also need to understand that many past assumptions about some of the most basic aspects of our lives, especially about energy, no longer apply.

In 2007, I wrote a book called *The Citizen-Powered Energy Handbook: Community Solutions to a Global Crisis* that focused on the more technical aspects of available energy technologies and provided some examples of those technologies at the community level. *Power from the People* picks up where that book left off, with a broader look at energy and conservation issues for individuals and communities as well as a more detailed and current description of how to organize, structure, finance, and launch local energy projects in your community.

Power from the People will help you and your community prepare for the emerging energy crisis. The strategies and models described are based on current, available technology with a strong emphasis on collaborative community initiatives based on local ownership. Local ownership is the key ingredient that transforms what would otherwise be just another corporate energy project into an engine for local economic development and greater energy security. Throughout this book, I will be highlighting various things that individuals, communities, businesses, institutions, nonprofits, and others around the United States are doing to build their energy resilience or otherwise curb their carbon footprint. Although the primary focus will be on the United States, we'll also take an occasional look at examples in other countries where some of the best models of community and local energy are found. Moreover, because thankfully there are so many energy efficiency and conservation programs in the U.S. today, these are not covered in depth here—see "Resources" at the end of this book and the companion website to this book series, resilience.org, for further resources on that topic.

The book is divided into four parts and fourteen chapters. Part 1 sets the stage, with a close look at energy that explains why our current reliance on non-renewable energy is unsustainable, and offers suggestions for better local alternatives.

Chapter 1, "Energy and Our Communities," offers an overview of our current main energy sources—oil, natural gas, coal, nuclear, hydropower, biomass—and explains how those energy sources are used to power our homes, businesses, and economy. It also shows why we are so vulnerable depending on twentieth-century energy (and assumptions) as we move deeper into a troubled twenty-first century where those assumptions no longer apply.

Chapter 2, "Conservation and Relocalization," takes a careful look at why we can't quickly switch everything over to renewables, and makes a strong case for conservation and relocalization. It explains which energy technologies lend themselves to local-scale projects and which don't, and why this is such a problem.

Chapter 3, "Rethinking Energy," looks at our use of energy from a "new normal" perspective and challenges conventional thinking about what we really need in our communities and where the energy that powers them will come from. It also looks at the many types of available renewable energy resources and how they can be harnessed for local use, both in your home and in your community.

Part 2 focuses on building individual energy resilience and offers guidance on strategies and programs that will help you to do it.

Chapter 4, "Your Household's Energy Resilience," explores *individual energy resilience* from a household perspective, now and in the future, including as it relates to location and transportation decisions. It looks at strategies for saving energy, beginning with energy efficiency and progressing to appropriate renewable energy systems to power and heat your home. It also describes programs for community-based initiatives.

Part 3 focuses on building community energy resilience.

Chapter 5, "Community Energy," offers the case for a *community energy* infrastructure and its many benefits as an alternative to a privatized energy industry with different goals. It covers various financial and legal structures that might be used to develop community energy, as well as the many barriers that often stand in their way, and takes an in-depth look at the technical structures (feed-in tariffs and the like) used in different locations to facilitate connecting local energy projects to the electrical grid. It also surveys various local municipal and cooperative organizing strategies for local energy projects.

Chapter 6, "Solar," begins the case studies. This chapter describes a community initiative in Colorado that is helping groups in cities and towns across the country organize their own community solar projects; a first-of-its-kind community solar project in Washington that continues to grow well beyond its organizers' original expectations; and a small community solar initiative at a cohousing project in Vermont.

Chapter 7, "Wind," profiles a small wind developer that is pioneering a community wind initiative in Washington based on a virtual net-metering strategy; a municipally owned community wind project in Minnesota that was able to successfully use a less-than-perfect site; and a cooperatively owned, commercial-scale wind farm for two small island communities off the coast of Maine.

Chapter 8, "Hydroelectricity," profiles a micro-turbine installation in Utah that sailed through the licensing process in just a few months; a small municipal project that took advantage of an opportunity to buy an existing privately owned hydro plant in South Dakota; and a very large hydroelectric project being constructed at an existing federal dam on the Ohio River for a consortium of municipalities.

Chapter 9, "Biomass," explores three biomass projects: a very successful, pellet-fired heating system for a small rural hospital in Oregon; a larger, more complicated community district heating system that relies on mill waste for fuel in British Columbia; and a much larger wood-chip-fired combined heat and power (CHP) project for a private Vermont college.

Chapter 10, "Biogas," looks at a locally owned start-up in Washington that recycles local farm and food waste into renewable electricity; one of the most ambitious municipal biogas initiatives in the nation, in Iowa; and a small, innovative project in Wisconsin that turns garbage into vehicle fuel.

Chapter 11, "Liquid Biofuels," focuses on a local biodiesel cooperative in North Carolina that is viewed by many as one of the most successful in the nation; a co-op in Massachusetts that has been impressively persistent and innovative in the construction of a local biodiesel plant; and a locally owned Iowa cooperative that is adding a cellulosic ethanol capability to its existing thirty-million-gallon facility.

Chapter 12, "Geothermal," profiles the oldest and largest municipal geothermal district heating system in the United States, located in Idaho; a combined heat and power geothermal system at a university campus in Oregon, the first of its kind in the nation; and hundreds of ground-source (geoexchange) systems used in a large, nonprofit community development located in Oklahoma.

Chapter 13, "Exceptional Community Energy Initiatives," explores a variety of comprehensive community-based projects and initiatives in Florida, Massachusetts, and Germany that stand out from all the others.

Part 4 is a call to action.

Chapter 14, "Be Prepared," offers a final overview of the challenges ahead as well as the opportunities and possibilities of adaptive resilience for you and your community.

An extensive guide to organizations and online resources, a glossary of terms, and a bibliography round out the volume.

Many of the strategies described in this book can be implemented by you in your own home (or even help you decide where your home should be located and what its energy needs and performance should look like). Other strategies we'll look at are more appropriate to help neighborhoods, communities, businesses, institutions, nonprofits, or even cities adapt to the "new normal." These strategies or projects can help move us from our present near-total dependency on increasingly expensive and fragile centralized energy to a greater reliance on local renewables in a downsized, relocalized economy.

However, we also need to be realistic. We use so much energy that none of these strategies can by itself replace our current demand. We must focus first on efficiency and conservation, and use the various strategies outlined in this book as a general guide. But if we use them in combination in the ways best suited to the unique characteristics of each community, we can significantly shift our energy future to be more resilient, more local, more environmentally friendly, and more democratic.

Opportunities in the New Economy

The business-as-usual, pre-2008 economy—based on endless growth of fossil fuels and consumption—is a thing of the past. The new economy is unquestionably going to involve significant reductions in consumption at all levels, and substantial changes to our lifestyles and patterns of living. Having said that, adapting to a new lifestyle does not have to be difficult or stressful. You can start with small, relatively inexpensive steps that begin your journey to greater energy independence and resilience at home and in your community.

This can be an evolutionary process of accepting that the future will be different, gracefully adapting your lifestyle, and adjusting your expectations to reflect that new reality. Happily, this process can be both rewarding and fulfilling. While it might involve reducing your present "standard of living," it can also result in a much higher quality of life for you and your family. And while you are on your journey toward becoming more energy self-sufficient, you will also discover the importance and many benefits of resilience at the community level. That's because you are only as secure as your neighbors.

Working with your friends and neighbors to come up with local solutions to global sustainability problems can be a challenging (and sometimes frustrating) but highly rewarding enterprise. I speak from personal experience. In 2005, I was a cofounder of the Addison County Relocalization Network (ACORN) in Vermont. In 2008, that group's energy committee was spun off and established a separate legal entity known as the Acorn Renewable Energy Co-op. The co-op provides education, outreach, and products and services to help its members make the transition from our present reliance on fossil fuels to greater use of renewables and local solutions. This has unquestionably been one of the most challenging but satisfying activities that I have ever been involved in.

If you get involved in a similar initiative in your community, like me, you will develop deep connections with like-minded people as you work together on projects and strategies to reduce your energy consumption while building more local energy resilience. This won't happen overnight, and it might not happen at all unless you get involved. Because, in the end, this is all about building community, one person and one project at a time. Your community needs you. And you need your community.

PART ONE

SETTING
THE STAGE

I

Energy and Our Communities

On a hot July day in 2003, I was sitting with a number of friends and acquaintances at SolarFest, an annual renewable energy and sustainable living fair held at the time at Green Mountain College in Poultney, Vermont.[1] One or two in the group were small-scale biodiesel producers, but all of us were concerned about peak oil and climate change, about our nation's addiction to fossil fuels in general and to oil in particular. We were bemoaning the lack of any organized liquid biofuels initiatives in the state at the time, and after a lot of complaining about the complacency of state government, someone said, "Well, why don't we do something about it *ourselves*?" This was a radically different view of energy policy, which up to that moment we assumed was someone else's responsibility. At the time, I don't think any of us fully understood what we were getting ourselves into.

Nevertheless, two months later about twenty-five people gathered in the basement of a church in nearby Middlebury for our first meeting about "doing something" to start a biofuels organization in Vermont. Among the participants were farmers, engineers, and biodiesel producers and users; Middlebury College faculty, staff, and students; several local legislative representatives; and interested members of the local community. Although the initial focus was local, we soon recognized that it needed to expand to include the entire state (Vermont is so small that even a statewide initiative is still basically local).

After several additional meetings and a lot of discussion, the Vermont Biofuels Association (VBA) was officially organized in November 2003. I was initially a co-director of the group and later served as president of the board. After five years of actively promoting biofuels in Vermont, the VBA was merged into a larger statewide group known as Renewable Energy Vermont (revermont.org), where some of its initiatives continue to this day.

The VBA was my first venture into local community-based energy activism, and involved a lot of rethinking of assumptions about energy that were current at the time—assumptions that left Vermont communities (and the entire state) extremely vulnerable to the ups and downs of international oil markets. One thing I quickly learned was that change comes slowly. But I also learned that once you stop waiting for someone else to "do something," and begin to think of energy in local terms, it opens the way for a wide range of possibilities and opportunities.

We're Vulnerable

We have a problem. The Great Recession has demonstrated just how vulnerable we are to disruptions in the global economy, and especially in global financial and commodity markets. Markets hate uncertainty, and the ongoing unrest in the Middle East and elsewhere has also shown how uncertainty can create havoc in oil markets—and increase price volatility at the gas pump.

U.S. oil consumption amounts to about 19 million barrels (798 million gallons) per day, and two-thirds of that is for transportation. Higher oil prices unquestionably played a role in the Great Recession, both in triggering the recession[2] and in slowing the post-2008 economic recovery—stretching many household budgets past the breaking point. Long term, the declining supply and rising cost of gasoline, diesel, and aviation fuels will affect the price of almost everything due to higher production and transportation costs. Oil is the lifeblood of the transport sector, the transport sector is vital to the global economy, and there are no obvious, viable alternative fuels (including ethanol and biodiesel) that would come close to meeting current consumption. Think about that.

Without oil, due to restricted supply or unaffordable prices (or both), our current global and national economies will eventually falter, leaving communities everywhere stranded without the basic necessities for daily life. That's because we have increasingly come to depend on faraway supplies for most of those necessities. Communities that do not prepare for this eventuality will have few alternatives, especially in the post-2008 world we now live in where many prior assumptions are no longer valid. Unfortunately, oil is not our only vulnerability. In order to better understand our predicament, we need to take a closer look at our current energy usage to see why our problems are not just restricted to oil.

Current Energy Needs

Most people in developed nations tend to take energy for granted. Flip the switch, the lights go on. Turn the thermostat up, and your home gets warmer. Press a button, and the microwave heats your meal. Start your car, and drive to the supermarket. But beyond that, many folks tend to be a little vague about the details of that energy—how it's produced, and how it gets to us. These are important issues, especially as the price of energy escalates and as supplies of non-renewable energy resources decline around the world. We are entering an era of what has been described by Michael Klare as extreme energy, due to the rising financial and environmental costs associated with most of the world's present energy sources. But in order to understand why this is so important—and to figure out where we need to go in the future—we need to know where we stand with our energy use today.

Our three biggest uses of energy in the United States are for transportation, electricity generation, and space heating and cooling—and all three are largely powered by fossil fuels. The transportation sector is the most dependent on fossil fuels, with more than 95 percent of all energy coming from oil.[3] Without oil, most people around the nation aren't going anywhere. In addition, the vast majority of our electricity (90 percent) is generated by fossil fuels and nuclear energy, neither of which is sustainable in the long term (hydropower accounts for 6 percent; other renewables and miscellaneous, the remaining 4 percent). It should be noted that electricity is a secondary source of energy; it stores and carries energy produced from other resources. Only 32 percent of the energy used to generate electricity actually reaches the end user; the other 68 percent is lost due to the inefficiency of conversion from coal or natural gas and from losses in the extended transmission and delivery system.

For home heating in the United States, natural gas accounts for approximately 54 percent, electricity 30 percent, fuel oil 7 percent, propane 5 percent, and "other" 4 percent. This means that around 92 percent of home heating in this country relies on unsustainable sources (I've given electricity a few

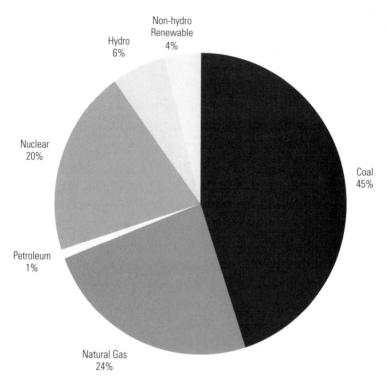

FIGURE 1-1. Total U.S. Net Electricity Generation by Energy Source (all sectors).
As of 2009, U.S. electricity sources included: coal (45%), natural gas (24%), nuclear (20%),
hydro (6%), other renewables (4%), and oil (1%). Data: Energy Information Administration, "EIA Electric
Power Monthly," U.S. Department of Energy, June 2011 (http://www.eia.gov/cneaf/electricity/epm/epm_sum.html).

percentage points since some of it is produced renewably). As you can see
from all these figures, our three main energy sectors are highly dependent on
sources that have no long-term future. You can also see that oil plays a very
small role in electricity generation, but a huge role in transportation and—in
some parts of the U.S.—heating fuels. The increase in heating oil prices in
recent years has left a lot of people—especially lower-income people—with
having to choose between paying for food and heating their homes.

So how does this relate to you? Most cars and light trucks need gasoline
or diesel fuel to operate. How do you get to work? How do your children
get to school? How do you go shopping for food and other basic supplies
for daily living? How do your mail and packages get delivered? Then there
is electricity. Your household appliances, including your heating and/or
cooling systems, need electricity to operate. Electricity makes just about

everything else in our technological society function. When it fails, virtually everything—lighting, water pumps, refrigerators, freezers, computers, cell phones, elevators, supermarket checkout lines—stops working. And without electricity and/or fuel (oil, propane, or natural gas) your heating/cooling appliances won't work. Consequently, you, your family, and everyone else is at risk if there is a steep rise in price or interruption of supply.

Energy in the Twenty-First Century

As we move into the uncertainty of extreme energy in the twenty-first century, we need to take a closer look at the individual energy sources that we depend upon for just about everything. Prices are going to rise (if not in real terms, then as a percentage of GDP), and supplies may be disrupted for a wide range of reasons. The peaking of fossil fuel extraction is right at the top of the list of concerns.

Peak Fossil Fuels

For many years, there has been a good deal of discussion—and disagreement— over the remaining reserves of fossil fuels. There is, however, no disagreement over the fact that these fuels—oil, natural gas, and coal—are finite and will eventually become so scarce that they are uneconomical to produce further. The only remaining questions are when (not if) they will enter permanent decline, and how quickly they will decline. Oil is right at the top of the list.

OIL

Global oil discoveries peaked in the late 1960s, and since the mid-1980s we have been finding less oil than we have been consuming. A peak in the discovery cycle always precedes the peak in the production cycle. Over the last decade, many petroleum geologists warned of the reality of an imminent "peak" in global oil production, and many concerned citizens, organizations, and policy makers have warned of the potential economic and social consequences of such a peak. Their predictions of tighter oil supplies and higher and more vola- tile prices have come to pass, especially with the slow recovery of the global economy in the years following the Great Recession of 2008. It is now generally accepted that we have already passed the peak of conventional oil production, and are bumping along the rough top of the bell curve of global oil extraction before we begin to slide down the long descent of decreased production.

Much of the world's supply of high-quality, light sweet crude oil (less than 0.5 percent sulfur) has already been exploited from "supergiant" oil fields that are now in decline. What's left are lower-quality and/or smaller deposits of conventional crude that are harder to find, more difficult to extract (think of deepwater and Arctic drilling), and more expensive to refine. In the years ahead, we will be depending more and more on "unconventional oil" such as heavy oil, oil sands, shale oil, coal-to-liquids, and so on. All of these sources are even more difficult and expensive to extract and process—to say nothing of the enormous environmental costs involved.

Combine this with continued demand growth in the developing economies of populous countries like China and India, and the stage is set for higher oil prices and increasing competition for dwindling reserves. When supply can no longer increase to meet business-as-usual demand, we reach a critical tipping point at which oil prices—and thus energy prices and the economy overall—start to become extremely volatile. This is exactly what happened in 2008 when oil surged from $100 to record highs of nearly $150 per barrel in about six months, and then plunged to $30 four months later. With the era of easy, cheap oil over, we can expect more of this roller-coaster ride in the years to come. The long-term prospects for oil are not good.

It's possible to create synthetic petroleum products from coal (as well as natural gas, oil shale, and biomass) using what is known as the Fischer-Tropsch process, developed in Germany in the 1920s. Unfortunately, production costs are high, and there are serious questions about the quantity and quality of coal supplies, not to mention the increased greenhouse gases and other pollutants that arise from using coal (more on a renewable use of the Fischer-Tropsch process with biomass later). Global synthetic fuel production capacity is around 240,000 barrels a day, a relative drop in the bucket compared with global oil consumption of 89 million barrels a day.[4]

Natural Gas

Although it is viewed as a cleaner (at the burner tip) fuel than oil or coal, natural gas is a non-renewable resource. In 2010, the United States consumed a record 66.1 billion cubic feet of natural gas per day. Natural gas is not only vital for space heating (54 percent) and electricity generation (24 percent), but also is used to produce nitrogen fertilizer for agriculture as well as plastics, fabrics, packaging, and many other products. Production of conventional natural gas in the U.S. has been in decline for a number of years, and until very recently we've been making up for this ongoing shortfall by imports—mainly from Canada and to a lesser extent by imports of liquefied natural gas (LNG) from overseas.

The gas from Canada is transported fairly easily via pipelines. But the infrastructure required for LNG imports—pipelines, terminals, double-hulled cryogenic tankers, and regasification terminals—is very expensive, controversial, and potentially extremely dangerous in the event of an explosion caused by an accident or deliberate attack. Moreover, the risks involved with relying on another imported supply of fossil fuel from potentially unstable or unfriendly nations are obvious.

In the past few years, there has been a lot of publicity about how natural gas will come to our rescue as a cleaner, climate-friendly "bridge fuel" that will help smooth our transition from coal and oil to renewables. Recent improved drilling and recovery technologies have led to the exploitation of shale gas deposits in the United States, and production has become fairly significant, accounting for around 23 percent of total U.S. gas production. This has led to wildly over-optimistic projections about how much of this gas is actually available and how much oil and coal it might displace in the next few decades while we shift to renewables.

Unfortunately, shale gas requires the use of high-cost, rapidly depleting wells that rely on huge energy and water inputs for a controversial process known as hydraulic fracturing ("fracking"). This process has been blamed for the pollution of groundwater in some locations, and the production of large amounts of toxic drilling waste. Additionally, a Columbia University seismologist has said that the pumping of huge quantities of drilling waste liquids underground "almost certainly" triggered eleven recent minor earthquakes around Youngstown, Ohio[5]—and a growing number of studies in the United States and United Kingdom have shown a correlation between fracking and local earthquakes. There is increasing opposition to fracking and shale gas exploitation in many locations around the country. Meanwhile, a growing number of individuals, including some geologists within the natural gas industry itself, are raising serious doubts about the hype coming from shale gas boosters.[6] Of course, the recent hype notwithstanding, natural gas supplies are ultimately finite.

COAL

The United States has been described as the Saudi Arabia of coal: It controls about 29 percent of global resources. In 2010, the U.S. consumed 1.05 billion short tons of coal, more than 92 percent of which was burned to generate electricity. But using coal—the dirtiest of the fossil fuels—in a way that is not environmentally disastrous is a real problem. Emissions from coal-fired electricity generation plants typically include carbon dioxide, sulfur dioxide, nitrogen oxide, carbon monoxide, volatile organic compounds (which form ozone), mercury, arsenic, and lead. There is no such thing as "clean coal," as its supporters refer to some technologies intended to reduce or clean up its emissions. These new technologies cannot economically reduce significant amounts of coal pollution, and they do nothing to avoid the mining of coal, especially the destructive mountaintop removal strategies used in the Appalachian Mountains that have devastated entire communities, wrecked local economies, and destroyed ecosystems. But it's the effect of coal's carbon dioxide emissions on the climate that has most scientists so worried. NASA climatologist James Hansen has described coal as "the single greatest threat to civilization and all life on our planet."[7] There is simply no way to make the coal industry clean or sustainable.

While construction of new coal-fired power plants in the United States has stalled over pollution concerns, China is said to have been building around one new plant *every week* in recent years. Despite the fact that about 60 percent of these new plants are more efficient and cleaner burning than those in the United States, China now burns more coal than the U.S., Europe, and Japan combined, making it the largest emitter of greenhouse gases on the planet.[8]

Several recent studies have raised the issue of "peak coal," the point at which global coal production will begin its inevitable decline. This is predicted to occur anywhere between the next few years and the next several hundred, depending on whose projections you believe.[9] But there is growing evidence that high-quality accessible coal reserves may run out much sooner than generally expected. That, coupled with increasing demand, especially from China, could hasten peak coal's arrival dramatically.[10] If these near-term predictions turn out to be the case, conventional wisdom about coal being a fuel for the long haul will prove to be just more wishful thinking. Since coal is burned mainly to produce electricity, alternatives to coal also need to produce electricity.

Alternatives to Fossil Fuels

Okay, so fossil fuels are problematic and will become increasingly expensive financially and environmentally in the years ahead. What are the alternatives? There are quite a few, but we'll start with nuclear power because it is viewed by many people as a solution to our global warming dilemma. Since nuclear power is used to generate electricity, it's primarily a replacement for coal and natural gas rather than oil. Let's take a closer look.

NUCLEAR

Nuclear power presently accounts for 20 percent of electricity generation in this country. One of the main attractions of a nuclear power plant is its ability to produce constant "baseload" power. In recent years, the nuclear industry has been busily promoting itself as a safe, clean alternative to fossil fuels. It's neither. One of the biggest problems with the nuclear power industry— which, after more than fifty years, has yet to be resolved—is what to do with all the highly dangerous radioactive waste that is generated. It appears that the Yucca Mountain nuclear waste dump in Nevada will probably never be used for its intended purpose. But even if it had been, the nation's existing nuclear waste already exceeds Yucca Mountain's original planned capacity.[11]

With nowhere to send it, the waste has been accumulating at the nation's 104 operating nuclear reactors at a combined rate of about twenty-two hundred metric tons a year. The nation now has more than seventy-five thousand tons of waste, overwhelming the on-site, water-filled cooling pools that were originally designed as temporary storage. The rest of the spent fuel has been placed outside in what is called dry-cask storage. This strategy is also intended to be temporary and is only designed to last for about a hundred years before it has to be moved to a safer, long-term storage facility.[12] For how long? Well over ten thousand years. That's a *very* long time to try to ensure safe storage even under the best of circumstances.

There has been a lot of talk in recent years—especially among certain politicians, industry lobbyists, and even some environmental groups—about the need to build more nuclear power plants in the United States. It's important to understand that this would never happen if the nuclear industry were left to survive on its own in the free market, because the numbers just don't add up. But in 2005, these same politicians and lobbyists managed to push through $18.5 billion in loan guarantees for companies that build new nuclear reactors. And there have been recent proposals for an additional $36 billion for the program.[13] Some proposals for using nuclear power as part of a much larger series of carbon reduction strategies call for doubling the number of existing U.S. plants. This would be enormously expensive, probably costing $6 to $8 trillion, and would almost certainly result in higher electricity prices as well.[14]

But what about the safety of *existing* generators? Despite repeated assurances by the industry that their facilities are safe, the March 2011 Fukushima Daiichi nuclear disaster in Japan once again graphically demonstrated just how dangerous nuclear power really is. Three of that plant's reactors experienced total meltdowns, hydrogen explosions and fires destroyed much of three of the containment buildings, and the resultant releases of radioactive pollution are now seen as second only to the 1986 Chernobyl disaster.

As a result of the Fukushima disaster, there has been a significant change in public opinion in a number of countries about nuclear power. A growing number of Americans are also opposed to the construction of new plants and to the continued operation of the twenty-three Fukushima-style GE Mark I boiling water reactors in the United States. The fact that safety standards on older nuclear reactors have been lowered repeatedly by the U.S. Nuclear Regulatory Commission in order to keep these aging facilities in "compliance" with those weakened standards should be a matter of great concern.[15] An Associated Press investigation also recently revealed that radioactive tritium (a radioactive form of hydrogen) has leaked from three-quarters of U.S. commercial nuclear power sites, often into groundwater.[16]

And speaking of water, too much or too little of it is another potential hazard for many nuclear power plants, which depend on vast amounts of water to absorb the waste heat left over after generating electricity, and to remove reactor core heat in the event of an accident. Future climate-change-enhanced floods or droughts could put the continued operation of many nuclear power plants at risk.

While there have also been some recent proposals for smaller-scale, modular reactors that potentially could be mass-produced, any plans to move ahead on new nuclear power plant construction in the United States—while ignoring both serious waste disposal and plant safety issues—would be highly irresponsible. Absent major breakthroughs to resolve these issues, nuclear power is a dangerous gamble at best.

It's important to note that nuclear, like fossil fuels, is based on a large-scale, centralized, corporate-dominated industrial model that is not easily downsized to local scale or ownership. Renewables, on the other hand, can be. However, we need to recognize that there are no perfect "silver bullet" solutions. Renewables have both positive and negative impacts, but on balance the benefits tend to outweigh the disadvantages, especially if these strategies are used sensibly in ways that benefit the entire community.

SOLAR

The sun is our oldest renewable energy resource, and its energy is either directly or indirectly responsible for virtually all of the renewable energy resources that we will be looking at in this book.

Solar energy can be harnessed passively or actively. Passive solar involves the use of building design and orientation to make use of the sun's rays to heat (and cool) living spaces. It's a very old strategy dating back thousands of years, and one of the simplest and most effective. It's also best suited to new construction, but can be incorporated into some retrofit plans. Active solar systems make use of specially designed equipment mounted on or near your home or other buildings to generate hot water or electricity.

Solar hot water systems (SHW) have been around commercially in the United States since the 1920s and today represent a mature industry in terms of the technology. Many different designs are available. Solar photovoltaic (PV) systems generate electricity with the use of PV panels that come in an increasingly wide range of styles and appearances. Although solar PV has been around since the 1970s, there have been some dramatic technical advances as well as price decreases in recent years as the scale of production has increased to meet growing demand. In fact, the solar industry became the fastest-growing part of the energy sector in 2010.[17] Solar PV panels are becoming an increasingly common sight on residential, commercial, and municipal roofs all across the nation. Although manufacturing these active systems does use a lot of resources, the "fuel" needed—the sun's energy—is free and non-polluting. Solar projects use modular equipment that can easily be scaled up or down, depending on their intended purposes. Small-scale solar installations are an important part of increasingly popular distributed generation strategies. The long-term future prospects for local solar energy are excellent.

Wind

Even though wind has been surpassed recently by solar in terms of annual growth, cumulative global wind generation capacity increased to around two hundred gigawatts in 2010. That represents an impressive 25 percent increase over 2009, despite the Great Recession. Much of that growth was in Asia, and especially in China. Wind provided about 1.9 percent of the world's electricity in 2011 and is expected to provide around 9 percent by 2020.[18]

First developed in the 1890s, wind turbine technology has come a long way in terms of size and sophistication. Early models could barely generate enough electricity for a few homes. Today's giants, rated at around six to seven megawatts (MW), can power entire communities. One huge new offshore wind turbine being built in Norway is 533 feet tall, with a rotor diameter of 475 feet; rated at ten MW, it will be able to produce enough electricity to power two thousand homes.[19] An even larger turbine, with a fifteen-MW generating capacity (also being developed for offshore use), is under construction in Spain and expected to be completed by 2020.[20]

Recent developments in blade and turbine design allow some turbines to operate effectively in areas of lower average annual wind speeds, making them practical in more locations. Wind power does not produce air emissions, generate solid waste, or consume or pollute water. Although wind turbines do occasionally kill some birds and bats, this is less likely with the larger but slower-moving blades on most modern commercial installations. Large-scale wind farms do consume a lot of resources—steel in particular, as well as rare earth minerals for magnets—but the biggest negative issue for some people is the visual impact of large wind farms placed along ridgelines (or in some cases offshore). Other people simply view wind farms as a necessary part of the working landscape. Like solar, wind power can easily be scaled up or down to meet the particular needs of homeowners, businesses, or communities. Although wind turbines will not work well in all locations, the general outlook for wind power is excellent.

HYDROPOWER

Hydropower has been around for millennia. Originally, it used the energy of falling or flowing water to turn a waterwheel, providing hydromechanical energy to power grindstones and other machinery. Then, in the 1880s, water-power was harnessed to generate electricity from spinning turbines. Gradually, hydroelectric power replaced hydromechanical power in most locations. Many of these local hydropower facilities have been around for so long that they are just taken for granted, but they provide valuable, steady generation capacity. Hydropower does not produce any significant emissions or solid waste, but the dams associated with some plants require vast quantities of concrete and . other materials to construct, and once built can kill large numbers of fish, flood valleys, displace people and wildlife, and eventually silt up, ending their useful life. Smaller diversion (or run-of-river) hydropower installations that don't back up large quantities of water behind them have much lower nega-tive impacts and are generally viewed more favorably by many observers.

Most of the obvious large hydroelectric dam sites have already been devel-oped in the United States, and due to the strong resistance to them in this country it is unlikely that any more will be built. However, there is consider-able potential for additional development of existing flood control dams as well as some larger navigational dams. There is also a lot of opportunity for development (or redevelopment) of a large number of smaller sites. A U.S. Department of Energy study identified 130,000 sites that could produce about thirty thousand MW of electricity from small-scale hydropower, with at least a few in virtually every state.[21]

The technology for these small-scale projects is mature and very reliable (although some of it has not been easily available in the United States because up until recently there was little or no demand for it). Most small-scale hydro-electric projects, especially run-of-river sites, have minimal environmental impacts and, once installed, cost nothing for the "fuel" (flowing water) to operate. Hydroelectric projects can be scaled up or down to match the avail-able water resource and to meet local needs. The future potential for small local hydro in the U.S. is good.

BIOMASS

Sunlight makes plants grow, and the organic matter from these plants is known as biomass. This plant material can be used as a fuel or energy source. Early on, our ancestors learned to use one form of biomass—firewood—for warmth, to cook food, and for nighttime protection from wild animals. It's still used for heating and cooking in many countries.

Today biomass in the form of pellets or wood chips can be used as a fuel for space heating, and in various forms can also be used to generate electricity. Electricity generated from biomass is sometimes referred to as biopower. Most biopower facilities use what is known as the direct firing method, burning the biomass feedstock directly in a boiler, producing steam that spins a turbine, which generates electricity. Some coal-fired electricity generating stations use a strategy known as co-firing, which reduces the amount of emissions by mixing biomass with the coal. Yet another strategy is gasification, which converts biomass into a gas, which is then used as fuel in a gas turbine or internal combustion engine that powers an electric generator. Recently, this strategy has been incorporated into small modular biomass systems that are available in a wide range of sizes, from five kilowatts to five megawatts. These systems use the gasification process to generate electricity as well as waste heat, which can be captured and used for a variety of purposes. The flexibility of different biomass fuels and the ability to place the modular system where the fuel source is located, rather than having to transport the fuel to the system, is a very attractive feature in many locations.

With the exception of the small modular systems, biomass technology is mature and proven. While the burning of biomass does produce carbon dioxide and other pollutants, the CO_2 is emitted into what is called a closed loop, where the majority of the CO_2 is used by trees and other biomass plants to grow the next crop of feedstock. Recent studies have looked at various potential sources of biomass feedstock, including quick-growing grasses and trees grown on marginal agricultural lands, in addition to selective harvesting in existing forestlands. While there are limits to how much biomass can be harvested sustainably, carefully managed local programs have good potential.

BIOGAS

Biogas is a subcategory of biomass. Biogas, composed mainly of methane (60 percent) and carbon dioxide, comes from anaerobic bacteria during the process of decomposition of organic matter in the absence of oxygen. Electricity can be generated with methane gas produced from organic wastes in landfills, or methane digesters on farms. Biogas can also be used as a substitute for kerosene or firewood for cooking, particularly in developing nations. That methane can be used for the generation of electricity as well as the production of high-quality fertilizer is a key benefit of biogas. Farm- and community-based biogas systems offer a lot of potential for local projects, particularly in the United States, where this is still not a widely used strategy.

LIQUID BIOFUELS

Because we are so dependent on liquid fuels (gasoline, diesel fuel, heating oil, kerosene), any renewable substitute should have a ready market. The good news is that biomass can also be used to create liquid biofuels such as ethanol and biodiesel. The bad news is that trying to produce enough of these fuels to make up for even a small part of our massive consumption of oil is a real challenge that can have unintended consequences, such as fuel versus food concerns. These concerns are about using farmland for biofuels production instead of for food. In 2001, approximately 7 percent of the U.S. corn crop was used for ethanol. By 2011, that had risen to around 23 percent.[22]

Ethanol is made by converting the carbohydrate portion of biomass into sugar, which is then converted into ethanol in a fermentation process that is similar to brewing beer. Ethanol is the most widely used biofuel today, with current U.S. production of around 13 billion gallons annually based mainly on corn (about 4.7 percent of current petroleum consumption). However, in response to ongoing concerns about using so much corn for fuel, more ethanol is being produced from cellulosic biomass feedstocks (rather than corn), currently around 6.6 million gallons and growing. Almost every gas station in the United States offers a gas–ethanol blend called E10 (10 percent ethanol, 90 percent gasoline). Some offer higher-ethanol blends, up to 85 percent (E85).

The other main liquid biofuel, biodiesel, burst onto the commercial scene around 1999 (although its early development work took place in the 1970s). Biodiesel production in the United States grew dramatically from 500,000 gallons in 1999 to 112 million in 2005, and then peaked in 2008 at 691 million gallons, according to the National Biodiesel Board. In the post-2008 economy, production fell to 315 million gallons in 2010 due to the loss of a key federal tax incentive, turmoil in financial markets, and questions about biodiesel's green credentials. Production rebounded dramatically in 2011 to a new record of around 1.1 billion gallons due to reinstatement of the tax incentive by Congress, high petroleum prices, and advances on non-food feedstocks and refining technologies. Biodiesel is now available at over 2,865 retailers and distributors nationwide.

Biodiesel can be easily made from virtually any vegetable oil through a simple chemical process; it can also be made from recycled cooking oil or animal fats. The process is so simple, biodiesel can be made by virtually anyone—although the chemicals required are hazardous, and need to be handled carefully. There have also been some promising experiments with the use of algae as a biodiesel feedstock, although its commercialization remains elusive.

While ethanol and biodiesel are not perfect, their production can be scaled up or down to almost any size, and they do provide a limited alternative to gasoline and diesel (respectively) for essential services and agriculture. Both ethanol and biodiesel offer considerable opportunities for small-scale, local, cooperative production initiatives, especially at the small family farm level (for more on biodiesel, see *Biodiesel: Growing a New Energy Economy*, second edition, Greg Pahl, Chelsea Green, 2008).

Biomass can also be gasified to produce a synthesis gas composed primarily of hydrogen and carbon monoxide, also called woodgas, biosyngas, or producer gas. The hydrogen can be recovered from this syngas, or it can be

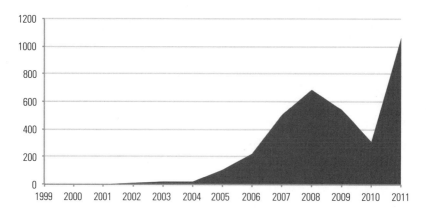

FIGURE 1-2. U.S. biodiesel production in millions of gallons. Data: National Biodiesel Board.

catalytically converted to methanol. It can also be converted using a Fischer-Tropsch catalyst into a liquid with properties similar to diesel fuel, known as Fischer-Tropsch diesel. Since it is based on biomass rather than coal or natural gas, this variant on the Fischer-Tropsch process is renewable.

GEOTHERMAL

Geothermal means "earth heat." Heat from the interior of the Earth is used to create more or less sustainable geothermal energy. Superheated groundwater or naturally occurring steam is tapped by drilled wells and used for the generation of electricity. Around 3,086 MW of electricity is presently generated from geothermal resources in the United States, mainly in California—and to a lesser extent in Nevada, Hawaii, and Utah. Some of these resources are also used for "district heating" of buildings (usually clustered in downtown locations) and sometimes for agricultural and industrial purposes. Since these geothermal resources eventually deplete, they are not strictly renewable, but some can last for several hundred years before new wells have to be drilled. The future potential for small-scale geothermal electricity generation is fairly good based on some recent developments in the technology that lower the cost and expand the range of the resource significantly.

In contrast, latent ground heat is not limited to geologically active regions, offering the opportunity for the use of what are known as geothermal exchange (geoexchange) heat pumps almost anywhere. Geoexchange heat pumps can tap into this resource and heat or cool a building. In the winter, heat is extracted from well water (or sometimes from fluid in pipes buried in the ground) and then increased with the help of a compressor. The heat is then distributed through the building, usually by a series of air ducts. In the summer, the process is reversed and the heat from the building is removed and dissipated into the ground. Geoexchange is the most energy-efficient, environmentally clean, and cost-effective space-conditioning system available, according to the Environmental Protection Agency (EPA).

Geoexchange heat pumps require electricity to operate, so unless you are powering them with PV panels, a wind turbine, or a small hydroelectric system, they are not entirely fossil-fuel-free. Nevertheless, geoexchange does provide a viable alternative to fuel oil and natural gas for heating and cooling. More and more homes, schools, and commercial and municipal buildings across the nation are being heated and cooled by geoexchange systems. The potential for even more local geoexchange heat pumps is excellent.

HYDROGEN

The "hydrogen economy" has been promoted by enthusiastic advocates as the answer to our energy dilemma. Hydrogen does have some very attractive attributes. It can be combusted like other fuels or it can be converted to electricity in fuel cells, emitting only heat and water as by-products. Fuel cells offer an alternative to gas-guzzling internal combustion engines for powering vehicles, and since hydrogen is the most abundant element in the universe, there is theoretically no problem with supply. However, many observers have their doubts.

It's important to understand that we cannot treat hydrogen as an energy source: There are no deposits of hydrogen gas that we can tap as we do with fossil fuels. Instead we must produce hydrogen from some other source, so it's really an energy storage and transport medium. Water is one of the most obvious sources; we can use electrolysis to separate the two hydrogen atoms from the one oxygen atom. Unfortunately, the electricity that would be required to produce the quantities of hydrogen envisioned would be enormous, far exceeding the present capacity of the national electrical grid (the nuclear power industry is quick to point out that they could provide that electricity—if they could just build more nuclear power plants). Trying to produce hydrogen from water is always going to be a net energy loser because of the first law of thermodynamics, which states: Energy can neither be created nor destroyed, it can only be converted between forms. And the conversion process is never 100 percent efficient.

Another major problem is the almost total lack of a national infrastructure to support a hydrogen economy. Trying to create that infrastructure would be enormously expensive, and would also divert precious resources from other promising technologies such as the direct use of electricity in electric vehicles, chemical batteries, and production of synthetic liquid fuels, which could offer similar benefits at much lower costs.

Last but not least, since the national hydrogen infrastructure (like the hydrogen economy itself) is mainly hypothetical, the fuel of choice for fuel cells today is natural gas or methanol. A fuel cell using these fuels produces carbon dioxide just like a conventional internal combustion engine. So what's the point? Many people now view the hydrogen economy as a Trojan horse for the fossil fuel and nuclear industries, which see it as a way to maintain their market dominance. While there may be some practical uses for hydrogen and fuel cells in some sectors of the economy in the years ahead, it simply is not the magic bullet solution that many people are expecting (or hoping) it to be.

Structural Challenges

Renewable energy strategies are clearly going to play a key role in the transition away from fossil fuels. But in order to make that transition, we will have to deal with a number of challenges. One of the biggest is the fact that most renewable energy technologies rely on fossil fuels for mining, transport, and production. Trying to reduce that reliance at a time when the supplies of those fossil fuels will be declining and their price rising will not be easy, especially since we are talking about decades for renewables to be able to scale up to significant levels to meet demand.[23]

The amount of time needed to bring a scientific "breakthrough" from the laboratory to commercialization is another key challenge. In general, it takes from twenty to twenty-five years for that process, meaning that even the most promising renewable energy development today won't achieve full-scale commercialization until the 2030s.

Ideally, it should be possible for a renewable energy technology to be a drop-in substitute for an existing non-renewable technology. Unfortunately, this is rarely the case; hydrogen, with its nonexistent infrastructure, is a prime example. Even wind and solar electricity on a large scale often require extensive investment in new transmission infrastructure to get the electricity from where it is generated to where it is needed (although as we will see later in this book, smaller-scale, community-based projects do not require expensive transmission infrastructure).

Resource limitations are another problem. Many advanced renewable energy technologies rely on rare earths and scarce metals whose supplies are finite and mainly located in China. This has not been a major problem— yet—but it may become a limitation as demand for these resources increases in the coming decades.

The intermittent nature of renewables like wind and solar is another challenge. Trying to figure out ways to store some of that intermittent energy for later use when demand calls for it is not easy. At the very least, we may need to adjust our expectations—and lifestyles—to match twenty-first-century reality. In addition, large quantities of water are used in the production of some renewables, especially some liquid biofuels, and future water shortages—to say nothing of crop failure due to increased climate instability—could have a serious impact on their production.

Last but not least, it's important to pay attention to the energy return on energy invested (EROEI) for all renewable (and conventional) energy strategies. Simply stated, if a process requires 1 Btu of energy to produce 3 Btu of biodiesel, the EROEI is 3 (expressed as 3:1). If the 1 Btu of energy results in less than 1 Btu of energy out, it's an energy sink or loss. The EROEI for hydrogen, for example, is around 0.5:1, making it an energy loser. There are many variables involved with producing most energy products, so EROEI is often quite variable as well: Conventional natural gas ranges from 15:1 to 20:1 and (non-"clean") coal ranges from 40:1 to 80:1, while wind power can range from 20:1 to 40:1. Despite the imprecise nature of these figures, the general ratios are useful when comparing one strategy with another.[24]

We need to keep all of these issues in mind as we try to transform our present energy economy into a more sustainable model. Relying more on local energy will help to mitigate at least some of these challenges.

Relationship to Other Issues

Climate Change

As mentioned earlier, the vast majority of climate scientists agree that the burning of fossil fuels is largely responsible for global warming, that the rate of warming is increasing, and that there's a connection to extreme weather events.

But it's not just the scientific community sounding the alarm. The conservative insurance industry has been studying extreme-weather-related losses for many years, and concurs with the scientific community on the dangers involved. "It would seem that the only plausible explanation for the rise in weather-related catastrophes is climate change. The view that weather extremes are more frequent and intense due to global warming coincides

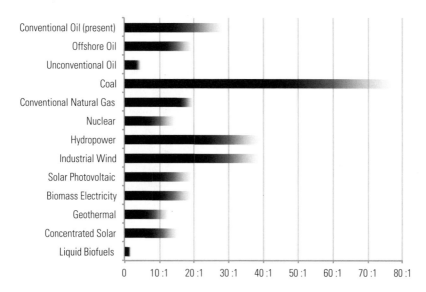

FIGURE 1-3. Energy return on energy invested (EROEI) for a range of energy sources.
Data: David Murphy, "A Tour of the Energy Terrain," in *Energy: Overdevelopment and the Delusion of Endless Growth,* eds. Tom Butler and George Wuerthner (Healdsburg, CA: Watershed Media, forthcoming).

with the current state of scientific knowledge . . . ," says Munich Re, one of the world's largest reinsurers.[25]

Unfortunately, despite all the talk about reducing it, more carbon dioxide was spewed into the atmosphere as a result of the burning of fossil fuels in 2010 than any previous year in human history, despite the downturn in the global economy. Meanwhile, politicians at home and abroad can't agree on what to do about it, while the climate change time bomb continues to tick.

The combination of extreme weather events in the United States and around the world has already strained the ability of governments and businesses to deal with them. Imagine these events occurring over and over again, with increasing severity in the years ahead. Disruptions to centralized energy delivery infrastructure—highways and especially the national electrical grid—will become more and more difficult and expensive to repair. The accumulation of the costs will almost certainly bankrupt some companies—even governments.

Geopolitical Instability
Our energy supply chains, especially for oil and liquefied natural gas, are vulnerable to disruptions in production and delivery caused by political

instability or natural disaster, especially in volatile regions such as the Middle East. The "Arab Spring" and "Arab Summer" of 2011 that spread from the Jasmine Revolution in Tunisia to Egypt, Libya, Jordan, Bahrain, Syria, and beyond caused political shock waves around the world—and helped to fuel a dramatic increase in the price of oil. This is yet another clear reminder of how vulnerable the United States is due to our addiction to fossil fuels in general, and oil in particular. China's insatiable demand for coal has helped to spur U.S. exports of coal in recent years. This raises the specter of domestic coal buyers having to compete in international markets. We need to shift our reliance to renewable energy as quickly as possible, and to local renewables as much as possible.

Economic Uncertainty

Energy markets in general, and the oil market in particular, hate uncertainty. The Great Recession and the subsequent economic, political, and social fallout have demonstrated just how interconnected—and fragile—the global economy has become. With one nation after another teetering on the brink of default on their debts, it should be clear that the global economy could collapse into chaos, leaving individual nations—and individuals—to fend for themselves. The global economy, national economies, and local economies were all built on twentieth-century assumptions about energy. Many of those assumptions no longer apply. We need to adapt to twenty-first-century reality, and the sooner the better.

Conservation and Relocalization

We've seen in chapter 1 that fossil fuels have no long-term future, and that there are some serious problems associated with our continued use of them. We've also seen that a broad array of renewable energy strategies are available as potential replacements. So why don't we just switch to renewables now and be done with it? There are four main challenges.

1. Fossil fuels, sometimes referred to as "fossilized sunshine," are extremely energy-dense. This means that we get significantly more energy out of them than we expend in finding, extracting, and processing them for our use. Unfortunately, most renewable alternatives are not as blessed with this trait, usually measured as energy return on energy invested. Consequently, replacing fossil fuels with renewable alternatives is not a simple one-for-one proposition in terms of energy content.

2. Then there is the issue of reliability. Renewable replacements for electricity generation such as wind and photovoltaics are intermittent and hard to predict since the wind does not always blow and the sun doesn't always shine (even during the day due to cloudy or rainy weather). To make up for this, sufficient baseload power must be available to cover the times when wind power and solar power aren't available; and most baseload power is currently provided by fossil fuels or nuclear power.

3. The sheer extent of our present dependence on fossil fuels means that making the switch to renewables will be an enormously expensive and time-consuming process. This also means that it will be virtually impossible to maintain our present levels of energy consumption with renewables, regardless of the technology used.

4. Last but not least, there are the Entrenched Big Fossil Fuel Interests. It's hard to understate how difficult it has been to make progress on renewables with the combined political and financial power of the fossil fuel industries, including many utilities that rely on coal or natural gas, fighting change at every step of the way for decades. In addition to using their political muscle inside the Beltway, Exxon-Mobil alone funneled nearly $16 million between 1998 and 2005 to a network of forty-three advocacy organizations that try to confuse the public on global warming science.[1]

Our massive petroleum dependency in the transportation sector presents some additional challenges. We've invested trillions of dollars over the last one hundred years into a transportation infrastructure built for gasoline and diesel—and that infrastructure is not easily (or cheaply) converted to run on, say, renewably generated electricity or hydrogen.

Moreover, we can't simply switch to oil-equivalent fuels such as biodiesel or ethanol. These two biofuels already consume vast quantities of food-based feedstocks (corn for ethanol, and mostly soybean, canola/rapeseed, and palm oils for biodiesel), with limited potential for further growth without greatly impacting food costs and availability. Some production growth is possible with nonfood feedstocks like switchgrass or algae, but even then ethanol and biodiesel production will almost certainly never reach nineteen million barrels per day, the amount of petroleum consumed in the United States on average in 2010 (largely for transportation). Finally, their low EROEIs preclude them from ever fully replacing even a major part of our current thirst for fossil fuels. This all means we need to drastically rethink our use of oil in general, and its use in our transportation sector in particular.

The huge multinational, corporate-dominated oil sector with its massive drilling and refining infrastructure as well as its sprawling supply and distribution chains does not lend itself to downsizing to the local community level. Considering all of these challenges, local communities need to think about providing some sort of alternative source of liquid transportation fuels in the years ahead.

Another problem is that there really are no other large-scale domestic alternatives to natural gas that will come close to meeting current or future demand. The huge natural gas production and distribution infrastructure in this country cannot easily be adapted for local use. However, as we shall see, there are small-scale biogas strategies available for use in local community projects. But again, the renewable options are definitely limited in the quantities of biogas they can provide.

Considering the huge role that coal plays in U.S. electricity generation, and the key role that electricity plays in the national economy, finding alternatives to this dirtiest of fossil fuels is going to be a real challenge. All of the renewables combined (excepting liquid biofuels, which are mainly for transportation) won't be able to replace coal anytime soon. There is simply no way to make the coal industry clean or sustainable. And there is no way to downsize the industry with its massive and sprawling infrastructure to a local, sustainable model either (not that this would make much sense in the first place).

What this all adds up to is a massive pattern of unsustainable energy use that is going to be extremely hard to change. At the very least, it will take many decades to shift to renewables at a significant scale, and even then they will never totally replace our current consumption levels of fossil fuels. Unfortunately, we don't have decades. This is why reducing our energy demand now is our best strategy.

Energy Efficiency and Conservation

Rather than building new electricity generation capacity, power transmission lines, and nuclear waste storage facilities to meet growing energy demand, it is faster and much less expensive to avoid and even reduce energy demand through energy-efficiency and conservation measures. Energy-efficiency measures generally aim to produce the same end result while using less energy, usually through improved technology; conservation measures, in contrast, reduce energy use by cutting the waste of energy, usually through a change in behavior or design. Both strategies reduce overall energy use, but achieve that goal in different ways.

Most states and utilities have recognized the benefits of energy efficiency and have been pursuing strategies to achieve them with a wide variety of initiatives in recent years. Targeting cost-effective low-hanging fruit, many utilities have encouraged their customers to install energy-efficient lighting and appliances, resulting in big electricity demand reductions in some locations: Targeted demand-side management programs in New York City and Westchester County trimmed thirty-six megawatts of electricity demand over a number of years.[2] Energy efficiency also benefits the environment by reducing the need for new power, decreasing the air pollution created by new fossil-fueled power plants. In support of Maryland's goal of a 15 percent per-capita reduction in energy consumption by 2015, Baltimore Gas and Electric Company customers have already achieved an overall energy savings of more than 390 million kWh, equivalent to eliminating the carbon dioxide emissions from the electricity use of nearly thirty-four thousand homes.[3]

Conservation measures "build in" lower energy demand through changes in behavior or design; for example, moving from a larger home to a smaller home, or moving from a car-dependent community to a community where walking, bicycling, and public transit are good options. Most people who pursue energy conservation measures to reduce their energy costs do so to save money, although environmental factors can be important motivators as well. Businesses generally tend to focus on conservation as a way to maximize profits. In either case, the end result is essentially the same: Less energy is used.

By reducing energy demand through efficiency and conservation, we can delay or eliminate the construction of new power lines, new pipelines, and new power plants. Energy efficiency and conservation should also be a first step for anyone thinking about retrofitting their home (see chapter 4). And both should be viewed as a first strategy to reducing carbon footprints while saving money at the same time.

But even the best of strategies can have its detractors, and energy efficiency is no exception. In economics, the Jevons paradox states that technological progress that increases the efficiency of the use of a resource tends to increase (rather than decrease) the rate of consumption of the resource. In 1865, the English economist William Stanley Jevons observed that the technological improvements that increased the efficiency of coal led to increased consumption of coal. In recent years, modern economists have studied so-called consumption rebound effects from improved energy efficiency. The main idea is that any energy savings are offset by increased energy consumption encouraged by the improved efficiency. Consequently, the Jevons paradox has been used by some, especially those who oppose clean energy, to argue that energy conservation is futile. This ignores the fact that a lot of people want to save money on their energy bills, and that energy efficiency offers a relatively painless way to do that. While there may be some debate—mainly in academic circles—about the rebound effect, the vast majority of energy experts affirm the value of energy-efficiency initiatives. There is virtually no debate about the multiple benefits of energy conservation.

Fossil fuel use will unquestionably decline if increased efficiency is accompanied by taxes that keep the cost of the fuel the same (or higher). Although it has been extremely unpopular in the U.S. Congress, the passage of a carbon tax should be a high national priority. Coupled with energy efficiency and conservation, these would make up a powerful set of tools to reduce consumption of fossil fuels and encourage renewables at the same time (two local carbon reduction tax initiatives have already been enacted in Boulder, Colorado, and the San Francisco Bay Area). However, big fossil fuel energy interests are firmly entrenched and not at all happy about losing their preferred status and treatment inside the Beltway, so don't hold your breath for much progress from Washington, DC. Unfortunately, our energy problems extend well beyond Congress.

As mentioned in the introduction, there are so many energy-efficiency and conservation programs in the United States today that they're beyond the scope of this book. See the resources section at the end of this book and the book series companion website resilience.org for information about groups working on these strategies.

The Structure of the Energy System

Centralized Energy

It should be fairly clear by now that our current centralized energy system will not be sustainable in an energy-constrained future. But this is not just about dwindling supplies of fossil fuels or increasingly unstable prices. High-profile scandals and putting profits ahead of consumer protection and the environment are ongoing issues with many big energy companies. The 2010 *Deepwater Horizon* oil rig explosion in the Gulf of Mexico, the 2010 Massey Energy Upper Big Branch coal mine disaster in West Virginia, and the 2011 Fukushima Daiichi nuclear disaster in Japan are prime recent examples. Lax regulation and conflicts of interest between industry and government have made a bad situation even worse. In all three of these disasters, an overly cozy relationship between the industries (oil, coal, and nuclear, respectively) and the government regulators who were supposed to be making sure things were operating safely was at least partly to blame for deadly emergencies.

This should come as no great surprise when you consider the sheer size of these corporate energy giants and the extent of their political influence. Virtually all of the big players in the energy sector have based their business models on large, complex, centralized systems that rely on interstate (and often international) supply and distribution chains of one kind or another to function. But these chains are only as strong as their weakest link. We need to keep this in mind as those chains and systems are tested again and again by a wide range of predictable (and unpredictable) stresses in the years to come. Eventually these chains will become more fragile and will begin to break.

This is why our continued sole reliance on interconnected, centralized energy systems is risky in the long run, and why finding some local alternatives for backup makes a lot of sense.

The Fragile Grid

With more than three hundred thousand miles of transmission lines, the national electric power grid is a huge and complex network that encompasses hundreds of companies and more than three hundred million users. But it's an odd and inefficient mixture of hundred-year-old design, fifty-year-old infrastructure, and some modern technology.[4] Basically, the grid is old, it's fragile, and it was never designed to be used the way it is presently. "We're trying to build a 21st-century electric marketplace on top of a 20th-century electric grid," says Ellen Vancko, a spokeswoman for the North American Electric Reliability Council. "No significant additions have been made to the grid in 20 years of bulk electric transmission, yet we've had significant increases in the amount of generation."[5] There have also been significant increases in consumption. Per-capita electricity consumption in the United States in 1960 was 4,050 kWh. As of 2008, it was up to 13,654 kWh. All of this combined has pushed the grid to its limits.

The August 2003 blackout in the Northeast and parts of the midwestern United States as well as Ontario that left fifty million people without electricity for up to three days was a glaring example of a "dumb" grid that didn't know what was happening—until it was too late—thanks to a cascading series of operator errors and equipment failures. And that's just a preview of what we can expect in the years to come unless we make some significant changes. Most solutions to this problem involve more large-scale, corporate projects that put even more strain on existing aging transmission lines or require huge new investments in more transmission capacity, or both. However, there is a relatively simple alternative: distributed generation.

Distributed Generation

Distributed generation refers to a power scheme in which the generation of electricity takes place not just (or possibly not at all) at a large centralized plant but in many smaller facilities distributed throughout an area—for example, solar photovoltaic panels on roofs. Such mini power plants might contribute to the main grid, or they may supplement grid-drawn electricity just for the households and business that own them. Distributed generation can reduce demand for out-of-state power during peak use (which saves a lot of money), help defer transmission line upgrades and expansions, and better insulate the grid from failure if a large, centrally located generation facility goes down. With emphasis on local electricity production, distributed generation can be far more resilient and efficient than centralized production, and can keep more dollars circulating in the local economy if the generation facilities are locally owned. Most of the discussion about local electricity in the rest of this book will focus on distributed generation.

The Smart Grid

Recent "smart grid" initiatives are a good foundation for future distributed generation projects. At a time when sophisticated computerized controls and information technologies have spread almost everywhere in industrialized nations, most of the U.S. electrical grid is stuck somewhere back in the 1950s. Maybe even farther back. The industry has generally been reluctant to make the needed investments in modern infrastructure for years. In many cases, utility workers wearing hard hats still have to drive trucks around to manu-ally shut off sections of the distribution grid with long, insulated poles. The situation on the consumer end hasn't been much better. Until fairly recently, there was no way for electricity consumers to get a grip on their daily electricity use and to manage it more effectively. Home-based technology is now available in the form of smart meters, smart appliances, and Internet-based programs that can provide real-time feedback on consumption, although it is only being used by a relatively small number of homeowners and utilities.

The smart grid is important for a number of reasons, and reliability is right at the top of the list. A smart grid would include sensors and other devices that would always know what is happening and would be able to automatically reroute power in an emergency to avoid future power blackouts. A smart grid would also allow various types of distributed generation to integrate more easily with the grid. This potentially empowers homeowners and communities to play a more active role in providing a larger share of their own power. It also helps their utilities to avoid having to buy expensive power on the spot market to meet demand peaks, which often involves importing power from somewhere else. But the main benefit for utilities will be the ability to balance supply and demand, leading to more rational power purchases when they are needed. This should also help utilities make better investment decisions.

However, there are some challenges and potential problems. Much of the existing electric industry workforce will need to be retrained in smart grid technology and systems. There are also concerns about the wireless technologies associated with smart appliances and their radio frequency emissions in the home as well as from large numbers of smart meters in neighborhoods. Some critics also raise privacy concerns while others worry about cyber threats. But probably the biggest challenge of all is the sheer magnitude of the electrical system transformation required.

Exporting Dollars

One of the biggest problems with most large-scale, centralized, corporate energy systems is that they tend to suck money out of the community. Currently, most energy dollars leave local communities and go to regional electric utilities or corporate or foreign suppliers of oil or natural gas. Once those dollars have been spent on importing energy into the community or state, they are no longer available to the local economy. This is a particular problem in the oil industry. Because U.S. demand far exceeds domestic supply, the oil sector has to import approximately half of the oil we consume—meaning that the majority of our oil dollars ($337 billion in 2010) are sent abroad, contributing to balance-of-payment woes and leaving less money for domestic needs.[6] But even the money paid for domestically produced oil leaves the community and does little to benefit local economies (especially in non-oil-producing states).

More and more states and local communities are trying to come up with ways to keep more of their energy dollars at home, and renewable energy investments are seen as one of the best ways to do that. But if communities are going to be playing a more active role in generating local energy, there is another important piece to the picture: local ownership.

Local Energy

Keeping your energy dollars circulating in your community is one of the biggest benefits of smaller-scale, local energy, and the key to that is local ownership. Local ownership of energy resources transforms what would otherwise be just another corporate energy project into an engine for local economic development. Instead of sending money out of state (or out of country), dollars spent on local energy projects have a multiplier effect—direct and indirect—in the community. The direct effect comes from the construction of the project itself, while the indirect effect relates to additional jobs and economic activity supplying goods and services to the project (as well as the profits retained in the community, if it is locally owned); this might also include local bank loans that keep local dollars circulating in the community. There is also an induced effect: the economic activity generated by re-spending the wages earned by those directly and indirectly involved in the project. All of this combined can add up to a significant economic benefit for local communities.

Investments in local renewable energy in particular help the local economy. These projects tend to be labor-intensive, so they generally involve more jobs per dollar invested (as much as three times more according to the Wisconsin Energy Bureau) than conventional energy projects. They also tend to use more local resources, so more energy dollars stay at home.[7] It's a win–win situation.

What types of local energy projects might work in your community? Most of the large, centralized, fossil-fueled energy systems for coal, oil, and natural gas that we rely on cannot be scaled down to local community size. Most cities and towns don't have oil wells or refineries in their backyards—to say nothing of coal mines. Virtually none of the sprawling fossil fuel infrastructure is adaptable for local use. Happily, many renewable energy systems can be scaled down for small community-sized projects:

- Electricity generation can be scaled down to the community and individual levels with solar, wind, and small hydropower, and to the community/regional level with biomass-fueled electricity generation and (in some locations) small-scale geothermal. On the downside, wind and solar only provide intermittent power, which can be unpredictable. Small hydropower, biomass, and geothermal electricity generation, however, can provide baseload power. Hydropower in particular can be extremely important in restoring power after a major grid failure. In the years ahead, if grid failures become more common, local electricity generation capability will be extremely important. Almost every community should have the potential for generating at least some of its own electricity.
- Liquid biofuels are vitally important because most other renewable energy strategies are not much help in the transportation sector. Biodiesel and ethanol can be produced fairly easily on a community or even individual scale in relatively small quantities (before they begin to compete with food crops for feedstock). Biodiesel produced from used cooking oil does not compete with virgin edible oils, but is becoming increasingly hard to find locally since its use as a biodiesel feedstock has surged dramatically in recent years. Free used cooking oil from your local fast-food outlet has disappeared in most locations. Nevertheless, liquid biofuels could play an important role for essential services in a new downsized, localized economy, especially in the absence of affordable petroleum.
- Biomass—firewood, pellets, and wood chips—can be scaled up or down to meet a wide variety of needs, from the individual home up to large industrial-sized projects. It can help replace some of our current use of natural gas, propane, and oil for space heating. And, as mentioned above, electricity from biomass is a viable strategy in some locations with sufficient biomass resources. When the heat produced by biomass to generate electricity is captured and used for

additional productive purposes (combined heat and power, or CHP), the process becomes much more efficient. Although some wood pellets are transported long distances, the main limitation on biomass is the local availability of feedstock, particularly for wood chips used in larger-scale heating and electricity generation. Because they are bulky and have a relatively low energy content (compared with fossil fuels), wood chips cannot economically be transported long distances. Biomass offers a lot of additional opportunity for local projects.

- Biogas facilities are particularly well suited to smaller local projects. They can be home-based to generate enough gas for a cookstove, farm-based to supply electricity to the farm, or industrial or institutional for several buildings or a campus. A biogas facility can be designed to provide both heat and power, offering the potential for greater efficiencies. Biogas sources can be landfills, wastewater treatment plants, and larger-sized food processing facilities. Biogas can also be used as a substitute for compressed natural gas (CNG) to run vehicles, although it must first be cleaned of impurities. Biogas can provide additional sources of income for farmers and rural communities, while reducing odor, pollution, and waste disposal costs at the same time. There is a lot of untapped potential for a wide range of local biogas initiatives.

Clearly, there are a lot of exciting local energy options. But developing local renewable energy does come with challenges: economic, organizational, political, and even environmental (more on this in the following chapters). It's also important to recognize that all of the available local energy sources combined will still not be able to replace our present massive consumption of fossil fuels, especially in the liquid transportation fuel sector. What this means is that, out of necessity, we will have to shift our thinking to a much more local economy and lifestyle—with a strong emphasis on conservation.

Relocalization

In the past, most energy was local. For centuries, waterwheels were used to power a wide variety of local mills and other manufacturing enterprises.

Beginning in the 1880s, hydropower was harnessed to generate electricity for those same mills and their surrounding communities. Then, in the twentieth century, most of those local community power companies were bought out by larger utilities, which eventually consolidated their holdings and closed many of the smaller power plants in favor of larger, "more efficient," central-ized power generating stations. The same general trends were followed in the other parts of the energy sector as well, leading to the centralized energy systems of today. But now, those trends are about to be reversed.

The risks involved in continuing to rely solely on these global sources of energy in the years ahead have been outlined above. Understanding the many benefits of local renewable energy, and the numerous options for producing that energy, stimulates a whole new line of creative thinking about energy—and the local economy. This is where "relocalization" comes in.

Local energy is part of the larger relocalization movement that more and more people around the world are actively supporting. This grassroots initia-tive involves a wide range of activities aimed at rebuilding more sustainable and more self-sufficient communities. Local energy has a key role to play in this. Communities that are working now to obtain more of their energy from local sources to provide power for their basic needs will be in a much stron-ger position than those that fail to do so. When fragile energy supply chains begin to fail, the primary concerns for most people will be food, clothing, shelter—and the means to provide those basic necessities. While it is possible to offer most of these necessities with hand labor, even limited amounts of local energy will make these tasks much easier. In the end, local energy production capability will be an important part of community resilience.

The coming shocks to the global economy from the age of extreme energy and climate change will force a major downsizing of the globalized distribu-tion system we know today. One of the main goals of relocalization is to build more resilient communities, which will be better able to survive these shocks, adapt quickly, and provide for as much of their needs from local and regional resources as possible. While local energy will make these vital tasks easier, we will have to rethink our present expectations of what that energy can do. We'll also have to rethink how we use energy at home and in our communities.

3

Rethinking Energy

As we move farther into the uncertainties of the twenty-first century, rethinking energy has shifted from an option to a necessity—and it involves taking a fresh look at how we produce and consume energy at every level, from the home to the community, region, and beyond. Critically, it also means rethinking how decisions are made about our energy system—and who makes those decisions.

Because most of the elected officials in Washington, DC, can't see past the next election cycle, it's become increasingly obvious that waiting for Congress or the president to "do something" productive about our fossil fuel addiction and climate change isn't enough. And the problem isn't just our elected representatives. National governments everywhere tend to be reactive rather than proactive, particularly with such crucial, complex issues. That's why getting organized at the local level is so important. Fortunately, we don't need permission from Congress to proactively begin addressing long-term energy issues from the ground up. And best of all, we can start immediately. This is an opportunity to rethink many of the basic assumptions about how our society and economy function from an energy standpoint. Those assumptions have led to many unsustainable patterns in our daily lives that we need to reassess before we can come up with an alternative plan that has a better chance for success in an extreme energy future.

Cheap fossil fuel energy has given us decades of economic growth, most visibly in the form of economic globalization: Most every U.S. household now enjoys the benefits of cheap imports like electronic gadgets from Asia, out-of-season fruit from South America, and trendy apparel from just about everywhere. But cheap energy has also had many negative consequences. Cheap oil spurred the phenomenal growth of interstate highways and automobile-dependent suburbs, and all those cheap foreign goods (sold in megastores accessible only by car) have decimated domestic manufacturing. Cheap coal and natural gas led to poorly designed housing (especially tract housing in suburbia) that is enormously inefficient to heat or cool.

If we start to seriously rethink energy, this leads to some pretty basic questions about other aspects of our "normal" lifestyles. Is it sensible to have a lifestyle that requires piloting a giant, expensive vehicle many miles every day to meet even the most basic needs of getting to work, acquiring food, and meeting up with friends and family? Does anyone truly *need* a five-thousand-square-foot, three-car-garage McMansion that requires hundreds of dollars' worth of natural gas (or oil, or electricity) every month just to keep it habitable?

Rethinking our homes and our personal lifestyles are good first steps, but they will only get us so far. If we're going to be serious about preparing for the era of extreme energy, we need to extend this rethinking of energy to our communities as well.

Local and Regional Energy

Fortunately, it's not necessary to live in a totally new community that has been designed for maximum energy efficiency in order to make productive use of local energy. There are many ways existing buildings and communities can be retrofitted for smarter energy consumption, and even production. Around the country, people and organizations are finding local energy solutions and innovations that save money, boost their local economies, and build local resilience. And as we'll see throughout this book, local ownership and local control are key to many of these efforts—both because that's how the benefits of local energy are kept in the community, and because it's often the local citizens, business owners, and government leaders who are most familiar with local resources, needs, preferences, and potential pitfalls. What you have to work with in terms of local energy in your own community may surprise you.

What Is Local Energy?
Until fairly recently, it has never occurred to many communities that they may have lots of local energy opportunities in their own backyards (or on their

rooftops). But when you start rethinking energy to look at ways of becoming more energy-resilient that don't necessarily rely on centralized, corporate-dominated utilities, a wide range of new possibilities begin to come into view. This has the potential to turn the usual "not in my backyard" arguments into "*please* in my backyard," especially if the community stands to benefit directly.

So what exactly is local energy? There are many possibilities. Simply stated, local energy projects rely on locally available renewable energy resources that serve local needs. One of the most obvious would be individual homeowner projects for solar hot water or PV-generated electricity that rely on the sun. That's about as local as it gets. But local energy can also include larger neighborhood, municipal, educational institution, cooperative, small-business, and commercial projects of many types. Another possibility might be a combination of these kinds of projects with, for example, a municipality working with a local cooperative (WindShare in Toronto, Canada, immediately comes to mind). There are many different possibilities, and there is no one "correct" approach. The best match between your available resources and your community's needs is the one to choose. If it's locally based and (preferably) locally owned, it qualifies as local energy. A community wind farm (or even just a single wind turbine) that is locally owned and generates electricity for the community qualifies. A large-scale, commercial wind farm owned by a distant corporate entity that sends its electricity—and profits—out of town does not really qualify.

Regional Energy

In some situations it makes sense to step back from local initiatives to look at how those individual projects might fit together in the regional picture. Not every community has a viable site for hydropower, but it might have a great site for wind energy. Another community might have excellent sites for large solar PV projects, but none for wind. A regional strategy allows communities to benefit from one another's strengths. It also means communities can share information, avoiding the need to reinvent the wheel for every project. A regional approach can especially make sense in an area where political boundaries artificially bisect a distinct energy-producing area such as a watershed or a mountain ridge.

In addition, some local renewable energy strategies are large enough to have a regional impact. Community biomass projects are a prime example. A number of communities in New England and New York have been studying possible district heating and power projects that are large enough to negatively impact forest resources in the surrounding region due to the potential for overharvesting. This has led a number of organizations collaborating on community-scale biomass projects to implement pre-feasibility studies, sustainable harvesting plans, and other services. The Regional Renewable Energy Advisory Network, based at the Northern Forest Center in Concord, New Hampshire, is made up of several dozen regional nonprofits, community groups, educational institutions, wood products industries, and state and federal agencies. The network has also collaborated on producing indexes, system diagrams, planning guides, financing forums, and roundtable discussions to advance greater sustainable use of available biomass resources.

In a separate initiative, the Nashua Regional Planning Commission in Merrimack, New Hampshire, has been assisting local community energy planning efforts for several years, according to Jill Longval, the group's environmental planner. "We started by helping communities do a basic inventory of their energy use to help them to see what they were spending on energy, how much energy they were using, what buildings were using more energy than others, and so on." More recently, the communities have started to use a regional approach to explore their renewable energy options. "There's a lot of research that goes into it, and communities can learn from each other when we take a regional approach," Longval says. "This has been a beneficial process to go through at a regional level rather than burdening each community with starting from scratch with all the research."[1] This strategy can make a lot of sense for many other communities and regions across the nation.

What Local Energy Resources Are Available?

We've already noted that electricity generation, liquid biofuels production, biomass, and biogas can be downsized to local scale. But how can these local energy resources actually be used by local communities and individuals to provide more energy resilience, and how do these resources stack up when compared with one another?

Solar

Although the sun is about ninety-three million miles from the Earth, solar energy is the ultimate local energy resource. Enough solar energy in the form of sunlight falls on the Earth in less than an hour to power the whole world for a year if we could capture 100 percent of it. Solar energy is non-polluting, free from wild price swings, and the supply is essentially unlimited. Best of all, solar energy can be "harvested" in many different ways for electricity, hot water, lighting, and space heating.

If you are lucky enough to be building a new home, you can incorporate a number of different solar energy strategies—especially passive solar—into its design. This can significantly reduce the amount of energy your home will consume. The same strategy works for municipal, commercial, or other building projects, resulting in energy-efficient, naturally lighted, comfortable spaces that can be a pleasure to work in.

However, most individuals and communities live in (or use) older buildings that were not originally designed with solar in mind. The energy resilience of these buildings can be significantly improved with new high-performance windows as well as active solar systems such as solar hot water (SHW), solar hot air (SHA), or photovoltaic (PV) panels for electricity. These SHW and PV systems are normally installed on roofs, but can be mounted on the ground under certain circumstances, offering a good deal of design flexibility (SHA panels are sometimes mounted on exterior walls).

Not all buildings can make use of solar energy due to their orientation, shading, or other factors. If that's the case, a relatively recent development known as group net metering (GNM) offers local individuals, groups, and municipalities some attractive new options for collaborative PV projects in locations that *do* have good solar exposure. Unless you live in a deep, narrow valley, or in a thick forest where there is little sunlight, you and your community probably have a lot of potential for greater use of solar power. The regulatory and approval process for SHW and PV installations is relatively quick and simple in most locations (more on GNM and regulations in chapter 5).

Wind

Many people would love to have a wind turbine in their backyard generating some or all of their home's electricity. Unfortunately, many of the best wind sites tend to be on ridges or mountaintops where most people don't live. This is where local community wind projects can be very effective. A community wind project offers a lot of flexibility in turbine location. You've heard the saying *Location, location, location* in the real estate business. With wind power, a good location is *really* important. With a good location (and a tall tower), a wind project is much more likely to be financially viable, so having a number of choices is a real plus.

Another advantage of local wind is that the technology is modular, meaning that you can match the size of the project to the need. If the power needs are modest, one turbine might be enough. If you need more electricity, add more turbines or increase their size. Assuming a good wind resource over a large area, the turbines can be placed where the power is needed, reducing the need for long-distance transmission lines. This is a key part of distributed generation strategies.

A local wind project can serve a particular municipal need, such as powering a sewage treatment plant, or can provide electricity for everyone. There are many possible ownership models (more on this in chapter 5). One other advantage of community wind is that, unlike most large commercial projects, a local community (or community group) might be willing to accept a lower financial return on their investment (perhaps due to a less-than-perfect location) as a trade-off for greater energy independence and resilience.

There is a lot of organized opposition to larger wind farms in many locations. Community wind projects, in contrast, tend to be smaller, community-owned, and directly beneficial to the community and local residents. This often changes the nature of the debate in favor of these projects. As one Dutch farmer and community wind supporter said, "Your own pigs don't stink." The regulatory and approval process for a wind project of any size, however, can be long, uncertain, and expensive in many states.

Hydropower

In the late nineteenth and early twentieth centuries, hydropower played a major role in local energy production. As late as 1940, hydropower represented about 35 percent of U.S. electricity production. Today that figure is around 6 percent. Nevertheless, hydropower offers some communities an opportunity to develop (or redevelop) small-scale, local hydroelectric sites that may have been abandoned. In addition, some sites that are still operating might be ready for upgrading with newer, more efficient equipment. In either case, hydropower's ability to generate electricity all day, every day makes it far more reliable than solar or wind power. Hydropower can be used to meet constant baseload or peak load needs, and it also plays a key role in restarting other power generation facilities (so-called black start capability) after a major power failure, making it even more attractive.

However, good hydropower sites are available only in some locations. Virtually all of the good, large (more than thirty-megawatt) hydropower sites in the United States have already been developed, and it's unlikely there will be any more (except in the case of some existing flood control or navigational dams that do not currently have hydroelectric plants, such as the Meldahl Project described in chapter 8). Most of the good small sites (one hundred kilowatts to thirty megawatts) have also been developed, though some of these were subsequently abandoned. But there are also numerous micro-hydro (up to one hundred kilowatts) sites on smaller rivers and streams that offer a lot of opportunities for local projects.

Any community that has a hydroelectric facility will have a reliable, relatively low-cost source of electricity regardless of the price or availability of fossil fuels. On the downside, the regulatory and approval process for hydropower is often long, uncertain, and very expensive, although there have been some recent efforts to streamline cumbersome state and federal processes for smaller projects. If your community has a good site, developing it will almost certainly be worth the time and effort required.

Biomass

Biomass is probably the most versatile of all forms of renewable energy. It can be used for space heating and also to generate electricity, and unlike solar and wind, biomass generation of electricity provides dependable baseload power, a real plus. Traditional biopower generating facilities tend to be relatively large and expensive propositions, but newer, smaller modular biomass systems may offer communities additional options.

Biomass can also be used to create liquid biofuels for transportation (such as ethanol and biodiesel) and a variety of other products, including chemicals and bioplastics, making it a partial substitute for oil. What's more, the feedstock for biomass is relatively local, offering the chance to reduce imports of foreign oil while keeping more energy dollars circulating in the local economy and creating long-term jobs in the growing, harvesting, and processing of these energy crops. But the main advantages of local biomass fuels to many people is their price competitiveness with increasingly expensive fossil fuels.

Firewood, our oldest form of biomass, has been used to heat homes for a very long time, and is still popular in many locations. In general, the firewood business tends to be small in scale and very local. The increased use of wood pellets for space heating in homes as well as municipal, state, commercial, and other buildings has spawned a growing number of wood pellet manufacturers, as well as stove, boiler, and furnace manufacturers and local dealers to sell and service these heating appliances.

While there are some concerns about just how much biomass can be harvested sustainably for wood pellets and wood chips, if forestry management guidelines are carefully followed, there should be a steady supply for the foreseeable future (admittedly, good management guidelines are not always followed in some locations). There is also some concern about potential competition between agricultural land used for biomass versus food production in some locations, as well as the possibility that increased biomass demand could encourage inappropriate salvage and thinning on publicly owned forest land.

There is some organized resistance to larger biopower projects, but the regulatory and approval process is fairly straightforward, particularly if there is strong local support. Increased reliance on biomass will strengthen agriculture, forestry, and rural economies in general, and should provide a greater degree of local energy resilience.

Biogas

When it comes to rethinking energy, biogas probably offers some of the most unusual and underutilized local energy possibilities of all. You might be surprised to learn that quite a lot of energy is generated underground in landfills, is sitting in manure lagoons on farms, and is even lurking in the entrails of cows that have been slaughtered for meat. Some of these potential resources may be located in your community—where they are perhaps viewed as a nuisance. They should be viewed as an opportunity.

Landfill gas—mostly methane—is regularly produced in landfills by the decomposition of organic matter. The Environmental Protection Agency requires that all large landfills install collection systems to prevent the gas from building up and causing an explosion and escaping into the atmosphere. An increasing number of landfills now capture this gas and use it to produce electricity. However, there are some concerns that increased demand for biogas could have unintended negative environmental consequences; for example, methane-capture facilities at landfills have created an incentive for some municipalities to favor waste disposal over recycling or composting, and may even release more methane (a potent greenhouse gas) into the atmosphere than would otherwise occur.[2] Moreover, biogas is admittedly not really a sustainable energy source, as it will eventually run out. But done carefully, it's possible to capture and utilize biogas in a way that makes environmental, economic, and energy sense.

A large number of European farms and a small (but growing) number of farms in the United States have been using methane digesters that convert cow manure into methane gas, which is then used to generate electricity. The solids remaining in the digester can be composted and turned into high-quality compost (or sometimes bedding for dairy cows). This process takes what is generally viewed as a management headache (manure) and turns it into useful electricity and a value-added product. This offers some interesting entrepreneurial opportunities for farmers (read about Farm Power Northwest in chapter 10).

The guts and other offal from slaughtered cows may seem like an unlikely energy source, but along with other organic waste, they have been used as feedstock for biogas to power municipal vehicles, buses, taxis, and even a local train in the city of Linköping in southern Sweden.[3] Admittedly, not everyone agrees with slaughtering animals for food, but this sort of imaginative thinking turns one business's waste into another's feedstock. It also offers some communities another opportunity to replace a portion of their dependence on fossil fuels with local power. There is little opposition to local biogas projects, and the regulatory and approval process is reasonably straightforward.

Liquid Biofuels
We use a lot of petroleum for transportation in this country—around sixty-three hundred gallons *per second*. Liquid biofuels are so important because our

current transportation system of cars, trucks, ships, airplanes, and their massive support infrastructure has been built to run almost entirely on liquid petroleum fuel. That system would take many decades to be even partially retrofitted for electricity, natural gas, or other energy sources,[4] so alternative fuels that can directly substitute for gasoline, diesel, and aviation fuel are essential.

Rethinking energy requires some particularly dramatic revisions to some of our most cherished assumptions about transportation. Petroleum fuels simply won't be available in the quantities or at the prices we have come to rely on for the last sixty years. The habit of hopping in the car just to pick up a gallon of milk at the grocery store five miles away will become a thing of the past.

While the term *biofuels* can refer to a number of different biomass-based fuels, it generally refers to liquid biofuels. The most widely used liquid biofuels are ethanol and biodiesel. Ethanol can be mixed with gasoline, while biodiesel can be used on its own or mixed with diesel fuel at various concentrations (biodiesel can also be mixed with No. 2 heating oil for space heating). Although there are substantial regulatory issues involved, both of these biofuels can be made in relatively small quantities at the local level by individuals, farmers, small companies, cooperatives, or other community groups. Making a profit with these small ventures, however, is not easy.

One of the biggest challenges for liquid biofuels has been price. When oil prices are low, biofuels have a hard time competing. When oil prices are high, interest in biofuels surges. This roller-coaster effect has made it difficult for many local biofuels producers to obtain long-term financing or to make a profit. But if the main goal of a local biofuel company or cooperative is to offer the local community some measure of security and resilience in transport fuels for basic community services, price and profit become less important, though they can't be ignored completely. The regulatory and approval process for local biofuels manufacturing facilities is considerable, but not insurmountable. There are reasonably good long-term opportunities for small-scale local biofuels.

Geothermal

To get a grasp of geothermal energy, just go outside and look down at the ground you're standing on: That ground can heat (or cool) your home.

That's admittedly a bit of an oversimplification, but that's the basic idea, and why it's so attractive.

There are two main types of geothermal energy that could be part of a community's local renewable energy resources. The first, high-temperature geothermal, is restricted mostly to geologically active locations, mainly on the West Coast, and to a lesser extent in Nevada, Utah, and Hawaii. There are also some geothermal resources located east of the Mississippi River in Arkansas, Florida, Georgia, North Carolina, and Virginia that are mainly associated with historic warm springs and spas.

The highest-temperature geothermal hot water and steam resources are generally used to generate electricity (geopower), while more moderate-temperature geothermal resources are often used for district heating. District heating can provide space heating for clustered buildings, mostly in downtown locations that make the installation of the heat distribution piping infrastructure more cost-effective. A fairly recent development in small-scale (around two-hundred-kilowatt) geothermal power plants capable of using moderate-temperature geothermal resources expands the potential range of geopower electricity generation for some smaller communities. One of the biggest advantages of geopower is that it can provide steady, baseload electricity around the clock.

The second type of geothermal heat, low temperature, is available almost anywhere. This normally involves extracting latent ground heat from the top fifteen feet of the soil (or from groundwater) and using it to heat buildings with the assistance of a geothermal exchange (geoexchange) heat pump. The process can be reversed in the summer to remove heat from the same build-ings. Geoexchange systems come in a wide variety of sizes and can be used by individual homeowners, municipalities, institutions, and other organiza-tions. The main disadvantage of geoexchange systems is that they use more electricity than conventional heating appliances (except for electric resistance heaters) and cost more to install. But over the life of the system, they can reduce heating costs by 30 to 70 percent and cooling costs by 20 to 50 percent.[5]

There are few if any regulatory or approval hurdles associated with individual geoexchange systems. Geothermal offers communities a lot of potential local energy options.

Using These Resources for Community Resilience

Because our addiction to fossil fuels is so pervasive, none of the local energy strategies we've just looked at will solve all of our energy problems. However, taken together as a part of a major rethinking of our energy needs, all of them (or at least those that are available in your area), if combined into a new local energy portfolio for your community, should offer at least some of the energy needed to help power basic needs in a new localized economy. It might not be enough for everything that we would want, but it might be just enough for what we need.

PART TWO

INDIVIDUAL ENERGY RESILIENCE

Your Household's Energy Resilience

Even if you live off the grid in a cabin, eat all of your own homegrown food, and don't have to commute to work, your household still depends on energy. If you live in a colder climate, you and your little cabin will probably need firewood to stay warm during the winter. And you will probably need some gasoline for your chain saw to cut the firewood, and perhaps for your pickup truck to haul the firewood. A PV-powered refrigerator might come in handy as well as a PV-powered well pump, not to mention a solar hot water system for your showers. And since your little cabin is probably located in the middle of nowhere, you'll need gasoline for the long drive to town to pick up the supplies that you can't quite live without yet—or for the occasional medical or other emergency. The point is that, however humble, virtually all households depend on energy to some extent.

Most of the rest of us, however, don't live in a remote, off-grid cabin, and we generally use a lot more energy to power our technology-dependent households and lifestyles. But this doesn't mean we can't be more energy-resilient. In this chapter, to get a better grasp of our overdependence on energy, we'll look at all the ways we use energy as individuals and families. Then we'll consider how the choices we make can have a profound influence on that energy use. The fact that greater energy resilience makes it easier to deal with energy shortages and higher prices is an added benefit.

How a Household Depends on Energy

If you've ever been through a power outage, you probably realize how dependent your home and family are on electricity. At the very least, your lights, refrigerator and freezer, microwave, food processor, radio, TV, computer(s), air conditioner, heater, dishwasher, clothes washer, and dryer won't work. If you rely on your own water well, the electric well pump won't work either. No lights, no heating or cooling, no water—that's enough to ruin almost anyone's day. And if you have a large freezer full of a year's supply of food, things can get a lot worse.

Even preparing the simplest of meals (to say nothing of cleaning up afterward) without water or electricity can be a real challenge. If you'd been planning on staying home and watching a movie, you might consider reading a book instead. But reading a book by candlelight or kerosene lamps isn't much fun either. You can probably forget about taking a hot bath or shower too. Many homes rely on natural gas for some of their appliances, but even some of these (especially central heating systems) won't work without electricity. When the power is eventually restored, resetting all of your digital clocks can be fun too.

You only have to go through this experience once or twice before you realize just how dependent your household is on energy in general, and on electricity in particular. If you live in a location that experiences frequent power outages, you may have already bought a backup generator. Chances are, though, that the generator runs on gasoline or diesel fuel—not exactly a long-term solution.

But even when the electricity is working properly, our households are extremely dependent on fossil fuels in other ways. Most pharmaceuticals and many health care and household cleaning products—to say nothing of pesticides and herbicides—contain a lot of petroleum-based chemicals. Countless household gadgets, toys, and games generally contain or are wholly made of plastics, as are most trash can liners, freezer bags, and food wrappers and containers. Most of the appliances we use contain a lot of plastic and are heavily dependent on fossil fuels in their manufacture. The list goes on and on.

Your household's energy dependence also extends well beyond your home. Unless you work in a home office, how do you get to work? People who live in cities generally have the option of using public transport. But if you live in a suburban or rural setting, chances are you probably have to drive your fossil-fueled vehicle—perhaps a long distance. How do your children get to school? Most ride on fossil-fueled school buses. How do their teachers get to school? Most drive. When you do your shopping at the supermarket (or perhaps food co-op), how do you get there and back? Fossil-fueled vehicle again? And if you forgot to put something important on your shopping list, you'll probably drive back for that too. When you go out to dinner or a movie or other form of entertainment, how do you get there? Taking a vacation? You're probably driving or flying.

How does your mail get delivered, and how do all those package delivery services get your latest online order from the supplier to your door? How do plumbers, heating and cooling system repair people, home repair contractors, telephone and cable service workers, fuel delivery vehicles, and building supply trucks get to your home when you need them? How do emergency services such as fire trucks, ambulances, and police arrive to help you? There's a lot of gasoline and diesel fuel involved in all of these activities, and if you live in suburbia you are particularly vulnerable. Our fossil fuel dependency is almost total for even the most basic functions of our daily lives in this country due to decades of bad planning, misguided public policy, and the false assumption that the fossil-fueled party would go on indefinitely. Although biofuels, bioplastics, and other renewables are increasingly available as substitutes, they will never completely cover our current, massive consumption of these products.

Happily, there are things you can do about many of these issues to strengthen your household's energy resilience. Some are relatively simple and can be done right away, while others may involve some long-range planning and rethinking some of the fundamentals of your daily life. In either case, it probably will involve making some big decisions today about where you will live and work and how that relates to your food supply, your family and friends, your recreational activities, and much more.

As with any decision about energy, the best first step is to identify how you can meet your needs using less energy. Start with conservation.

Energy and Transportation

The most important choice you have regarding energy resilience is where you live. That's because where you live has an enormous effect on your transportation needs, and your transportation needs have an enormous effect on your energy consumption.

Back in 1999, my wife and I were thinking about moving from a convenient (but noisy) in-town location to a new home in a less congested neighborhood. While we would have loved a quiet country location, we decided that our new home had to be within walking distance of her work. This was a conscious decision to avoid having to commute to work, but it limited our possible choices to a few neighborhoods and streets. As homes came on the market, we looked at each one, and decided to wait until the right home in the right location came along. We looked, and looked—and looked. Homes came and went. Finally, in late 2000, a house appeared on the market that met our criteria. It was a few blocks farther away from downtown than we had originally intended, but the daily walk to work would be good exercise. While we could have easily found a house out of town that otherwise would have met our needs, we decided that the wait was worth it. Not having to be car- and fossil-fuel-dependent to get to work was very important to both of us, and we've never regretted that decision.

At the time, I had not yet heard the phrase *energy resilience*, but that was exactly what we were strengthening by living near where we worked. This seemingly simple decision to live near your work could be one of the most dramatic ways of cutting your fossil fuel dependence and reducing your household's carbon footprint while enhancing your energy resilience. Win, win, win.

One of the simplest and least expensive transportation choices to improve your household's energy resilience is to rely on your own two feet—easy to do if you've chosen to live in a relatively compact community or mixed-use neighborhood.

Bicycling is also becoming increasingly popular. Even in large cities, more and more commuters are discovering that they can get to work, school, or the grocery store easily and safely on a bike, and get a good workout while doing so. You can use your own bike, or, in more than a dozen U.S. cities, you can take advantage of a bike-sharing program; worldwide there are between two and three hundred such programs, and more keep popping up all the time. Of course, bicycling requires a set of skills, habits, and equipment that are different from driving a car or using public transit. In hot weather, you might schedule time to shower and change clothes at a gym near your work; in rainy weather, you'll need a good set of rain gear and lights. But as the bicycling residents of the cold, snowy cities of Northern Europe know, if the infrastructure is good you can bike in just about any weather.

If you live in a city, you can make use of public transit, one of the most energy-efficient ways of moving large numbers of people from one place to another. In some cities, the public transit system is so good you really don't need to own a car. Unfortunately, public transit options are limited or nonexistent in many parts of the United States, particularly the suburbs and rural areas. In that case, purchasing a high-gas-mileage car may be the only option for improving your energy resilience in personal transport.

If you use your car mainly for commuting and you don't drive much more than forty miles one way, a new plug-in electric car might work for you. But if, like most people, you use your car for both short- and long-distance driving, one of the new plug-in electric hybrids offers a lot of flexibility as well as improved gas mileage. When you are just driving around town, you can rely on the battery pack. When you need to travel a longer distance, the gas engine takes over—no worries about your batteries running out in the middle of nowhere. While you will still be using some gasoline, it will be a lot less.

If you have no choice but to drive to get to work, school, and wherever else you need to go, some simple changes to your driving behavior can improve your energy resilience right away. Carpooling to work is an obvious strategy to at least spread the burden of fossil fuel dependence among a number of people. Admittedly, trying to get your schedule to mesh with others' can be a challenge, but removing at least one regular car trip from your lifestyle— even occasionally—is worth it. You can carpool every day or just on certain days, both directions or one way, depending on your needs and schedule. In addition to saving you money on gas and parking fees, it can also save wear and tear on your car—$3,000 or more per year in some cases.[1] Many states have ride-sharing websites to make carpooling and vanpooling easier. North Carolina's PART (Piedmont Authority for Regional Transportation) website (www.partnc.org) is one of many. For example, Edna Barker of Winston-Salem, North Carolina, was able to organize a vanpool with the assistance of PART. "If it were not for PART listing 'Share the Ride NC,'" she says, "I would never have known that there were ten other people within 10 miles of my home who were interested in sharing a ride."[2]

Another fairly simple change to driving behavior is "trip chaining." This involves a little planning in advance to combine a number of errands into one trip, and a little thought about the best route you should take to make it as efficient as possible. A shopping list might be handy too, so you don't forget anything, but that's pretty much it. This simple strategy could save you several hours a week and money at the gas pump.

Car sharing is another strategy that is attracting a growing number of participants. Originally started in Switzerland and Germany in the 1980s, the idea spread to Quebec City in Canada in 1994 and to Portland, Oregon, in 1998. Today, there are approximately twenty-seven car-sharing programs in the United States with nearly 520,000 members sharing around seventy-eight hundred vehicles, and seventeen programs with more than 85,000 members sharing twenty-four hundred vehicles in Canada.[3] Basically, it's a car rental service, but mainly for just a few hours rather than days. You only pay for how long you use the vehicle and—in some cases—the distance traveled. While car sharing does not eliminate driving, it can reduce the number of cars on the road; one shared car replaces at least eight individually owned cars. Car sharing is found in a growing number of cities, although rarely in rural locations. The largest car-sharing operator in North America is Zipcar, but there are many others. CarSharing.net is a good place to start for more information on this strategy.

Not always having a car of your own immediately available and having to consciously pay to use one will almost certainly change your driving habits.

Energy and the Home

There are two main strategies: reducing the amount of energy you use, and generating some of your own. Finding ways to reduce your energy consumption is the first step, because simply adding solar panels or other renewable energy systems to an energy-inefficient home is a waste of money and valuable resources. It's much more cost-effective to start with an energy-efficient home.

Saving Energy

What kind of home do you live in? This is an important question because it will help inform you where to place your emphasis. Look at your home, its construction, and its location. If you live in a cold climate zone, obviously your main concern is staying warm in the winter—as is staying cool in the summer in a warm climate zone. Is your home an old, drafty farmhouse with little insulation and single-pane windows, or is it a recently built home with a lot of insulation and tight, double- or triple-glazed windows already installed? What kind of heating and/or cooling system does your home have and how is it powered or fueled? The heating, ventilation, and air-conditioning (HVAC) system is the largest energy consumer in most homes. Every house is different: A strategy that might make sense in one home might be a complete waste of time in another. There are energy-efficiency professionals who can help you sort this out (more on them later).

After the HVAC system, your water heater, refrigerator, clothes dryer, and dishwasher are the next largest energy consumers in most homes (you can find out how much energy typical household appliances use at energysavers .gov). That's why one of the things that almost anyone can do to cut energy use is to replace older, less efficient major appliances with new, more efficient ones. Almost all major appliances have EnergyGuide labels that list approximate annual operating costs and help you compare energy consumption between different models of similar appliances. Televisions and most other audio and home entertainment appliances also now have these labels.

When it comes to smaller appliances, however, the situation is not so simple. This is because many smaller appliances manufactured in recent years—regardless of their energy efficiency—draw electricity even when they are turned off. In fact, nearly 20 percent of the electricity used by these appliances is wasted while they are in standby mode waiting to be used.[4] Almost any small electronic device that is plugged into a charger or outlet when it is not actually in use is probably wasting electricity. TVs, cable boxes, satellite dishes, and so on account for as much as 35 percent of this leaked electricity (also called phantom power loss). Audio equipment represents another 25 percent. Your stereo system may be wasting as much as 93 percent of the energy it consumes during a twenty-four-hour period—while it is switched off. The only way to stop this leakage is to unplug these appliances, or plug them into a switched power strip and turn the power strip off when you're not using the appliance. Small appliances now consume about 20 percent of total household electricity, and that number is increasing with the proliferation of rechargeable handheld gadgets.[5]

Some conservation strategies are quite basic and can save a lot of energy if you employ them on a regular basis. Many don't cost anything. For example, there are some things that can be done by hand or eliminated altogether, such as drying clothes on a line or a rack instead of with a dryer. About 5.8

percent of residential electricity use goes to clothes dryers, and 23 percent of clothes dryers in use are heated with natural gas; multiplied by the 87.5 million Americans who use clothes dryers in their homes, this can add up to quite a lot of energy.[6] While we're in the laundry room, you can also save energy by washing in cold water: About 90 percent of the energy used in a clothes washer goes to water heating.

Other simple behavior strategies that don't cost anything include adjusting your air-conditioner thermostat up and your furnace thermostat down. For every degree change in the setting you can save from 1 to 3 percent on your cooling and heating bills, respectively.[7] Using fans instead of the air conditioner can save even more. Always turn off unnecessary lighting. Enable the "power management" feature on your computer and always shut the computer off at night. If you use a dishwasher, use the air-dry setting instead of heat-dry (this simple strategy can cut its energy use from 15 to 50 percent), and only run the dishwasher when it's fully loaded.

With all this emphasis on saving energy, you may be wondering how you and your household's energy footprint compares with others. Check out the Carbon Footprint Calculator from the CoolClimate Network at the University of California, Berkeley (see http://coolclimate.berkeley.edu), which will also give you a rough sense of the energy your lifestyle requires. The calculator allows you to enter details about your household, transport, housing, food, goods, and services. At the end, there is a summary that shows how you stack up against similar U.S. households, average U.S. households, and a world average (which will probably be an eye-opener).

These days there are many resources and programs available that can help you reduce your energy consumption. One of the most popular is the Low Carbon Diet, based on the book of the same name by David Gershon (see www.empowermentinstitute.net/lcd).

Strategies for Homeowners

After taking the steps above, most homeowners aiming to further build their energy resilience opt for retrofitting their homes: with better insulation, with energy-generating investments such as solar panels, or even with redesigning whole portions of their houses for smarter energy use (good-bye cathedral ceiling, hello extra storage space!).

But as radical an option as it sounds, it may be worth considering moving to a more energy-efficient home closer to your work and the destinations you frequent (grocery, shops, place of worship). After all, if you are appalled by your household's overall energy footprint, it's likely that the *location* of your home—and its implications for your transportation behavior—played the biggest role after air travel. Or perhaps you don't want to sink a lot of money into your present home, or you're older and you're ready to downsize now that your kids are launched.

If you are going to move to a different home, regardless of its location, you will want to take a careful look at it from an energy-efficiency perspective. Ask a lot of questions about the home's HVAC system, insulation, windows, orientation to the sun, and especially the utility bills (admittedly not the first things on most home buyers' priority lists, which usually focus on views, schools, and kitchens). Huge homes with extra rooms should be off your list; ditto for homes located in the middle of nowhere (or far outer suburbia) with no access to public transit. Why trade one bad energy situation for another? Your new home may not be perfect, but it should be relatively energy-efficient, or at least upgradable without spending too much additional money on it (more on this in a moment).

The most radical route to go is to build a new home from scratch. If you do so (in a location with ample walking, bicycling, and public transit infrastructure, of course!), you should seriously consider a passive solar design that makes maximum use of the sun for natural heat and lighting. Building a new home without solar being the main part of the design makes no sense. It's not much more expensive to build for passive solar than with a conventional design; it simply uses the same building materials in slightly different—but important— ways. It also involves orienting the home to take the best advantage of the available solar resource, something that is rarely done. From an energy standpoint, almost all other considerations are secondary. A properly sited and designed passive solar home will dramatically cut your energy consumption, increase your comfort, and give you a naturally light, pleasant living space.

If buying or building a new energy-efficient home is not an option, or if you already live near where you work, a home-energy-efficiency retrofit might be your best option, especially if you live in an older home. Happily, this strategy also has the potential for offering the quickest return on your investment and should be the first step before you embark on installing any renewable energy system. Sealing air ducts or adding insulation that you can't see is admittedly not as sexy as installing highly visible renewable energy systems like solar PV and solar hot water, but most energy-efficiency experts will tell you to start with a retrofit. It will also significantly enhance your energy resilience, cut your utility bills, and increase the value of your home.

What's involved with a retrofit? There are many possible options, and it can be a bit daunting to decide where to start. Most residential energy-efficiency experts recommend that you begin with a home energy audit conducted by an experienced home performance contractor. The contractor will guide you through the most cost-effective things that can be done to help save the most energy. This can include replacing old appliances with new Energy Star–qualified models, switching to energy-saving lightbulbs, or more comprehensive energy retrofit projects that include sealing air leaks, adding attic and sidewall insulation, replacing your HVAC system or drafty old, single-pane windows, and so on.

The right strategy depends on your home and its construction details, where you live, the size of your budget, as well as your personal goals. Insulation and caulking may not be very exciting, but they might make more sense from a return-on-investment perspective than expensive new windows. This is why it's important to start with unbiased information from a nonprofit or utility-based residential energy conservation program in your state that's not trying to sell you a specific product. These programs will also help you find qualified professionals to do the retrofit work you decide on (see "State-Based Programs" in the resources section at the end of the book).

Generating Energy

Only after you've exhausted all reasonable options for reducing the energy your lifestyle and your home require does it make sense to invest in generating your own energy. In addition to identifying the kinds of energy applications you require most (in Minnesota it may be space heating, while in Florida it's probably electricity), be sure to also consider ease of repair and equipment life span. A simple solar hot water system is relatively easy to repair compared with a high-tech sun-tracking array of solar photovoltaics. Also, be sure to look into incentives and credits offered by your utility and state and federal governments.

Solar Heating

Solar energy can be harvested for a number of residential purposes. If your home is not a suitable candidate for a passive solar retrofit, an active solar heating system might work. In some locations (particularly in the South-west), heating your home (or at least some of it) with a solar hot air system is popular. However, heating your home with solar hot water is a far more popular strategy in most parts of the country. If you already have hydronic (circulating hot water) radiant slab floor heating in your home, adding a solar hot water space heating system can be an excellent strategy. But a simpler and less expensive approach is solar hot water (SHW). These systems produce the hot water you use to wash dishes and clothes, take showers or baths, and so on, but do not heat your home; that's why they are sometimes referred to as solar domestic hot water (SDHW) systems.

FIGURE 4-1. Solar hot water collector panels on the garage roof of the author's home.
Photo by Greg Pahl.

Since your water heater is usually the second largest energy-consuming appliance in the home (after your HVAC system), it makes a lot of sense to try to produce your own hot water with the sun. While the fuel (solar energy) for your SHW system is free, the hardware isn't. Nevertheless, when you factor in the various state and federal rebates and other incentives that are generally available, a SHW system can be fairly attractive from a financial standpoint—to say nothing of an energy-resilience point of view. There are many different SHW system designs. The type that is best for you depends mainly on what part of the country you live in. Cold climates require designs that avoid frozen pipes and collectors, while warmer climates don't.

Ideally you need a south-facing roof (or within thirty degrees of true south) with unobstructed solar exposure between 9:00 AM and 3:00 PM. But even if you don't have one, solar collectors can be ground-mounted, offering quite a lot of design flexibility. As long as you have these basics to work with, you have potential SHW options.

Most SHW systems generally cost somewhere between $10,000 and $15,000. A solar hot water home space heating system can easily cost $20,000 or more. There are numerous utility, state, and federal incentive or tax credits for residential solar hot water installations that can reduce these prices. Most communities require a building permit for the installation of a solar hot water heater onto an existing house, but the process is generally straightforward. Check with your local municipal officials first to be sure.

My wife and I have had a simple, PV-powered, two-panel SHW system mounted on our south-facing garage roof since 2004. Our household gets virtually all of its hot water in the summer, and some during the winter, from the system. We've been totally satisfied with it.

Solar PV

Of all the home-based renewable energy systems, solar photovoltaics (PV) get the most attention. There's nothing terribly glamorous about cellulose insulation gathering dust in your attic, but watching your electric meter spinning backward can be very satisfying (even though the attic insulation probably offers a much quicker payback).

A PV system uses a large number of photovoltaic cells on collector panels (or other special surfaces) to convert sunlight into direct current (DC) electricity.

A group of panels combined are called an array. In off-grid settings, the DC current from the array is fed into storage batteries for later use. However, in most grid-tied systems the DC current is converted into regular alternating current (AC) power with an inverter and fed directly into your home's electrical circuitry. Any excess power not being consumed by your home is fed back into the grid (causing your electric meter to spin backward, reducing your electric bill). This net-metering strategy is available in all but four states (Alabama, Mississippi, South Dakota, and Tennessee), although the specific details of how the programs work vary from state to state.

There have been a number of new second- and third-generation collector materials entering the market recently (including some that look like roofing shingles), which offer a lot of flexibility in appearance and installation options. However, PV panels are still relatively inefficient (around 20 percent) at converting solar energy into electricity, requiring a lot of panels to produce significant amounts of electricity.

The same general site rules about exposure to the sun as well as the orientation of your home for solar hot water also apply to PV systems. PV panels are normally roof-mounted, but wall or ground mounting is also possible. One added option to the ground-mounting strategy for solar PV is the ability to occasionally adjust the angle of your panels to match the seasonal angle of the sun with some mounting racks. Another (more expensive) option is a ground-mounted rack that allows the array to automatically track the sun during the course of the day. Both of these options increase the amount of electricity produced over fixed mounting. Add to all of this a number of other electronic components, and a solar PV system can easily cost $10,000 to $40,000 or more, depending on a wide range of variables. This is why it's so important to reduce your home's electricity consumption *before* you install a solar PV system.

Electrical (and sometimes building) permits are required by many municipalities to install PV systems on your home, and states or utilities normally oversee the net-metering interconnection process.

Backyard Wind

Many people, especially those who live in rural locations, have visions of a wind turbine spinning atop a tower in their backyard that provides all their electricity. There are several problems with this vision. The first is that unless you happen to be extremely lucky, most backyards have poor wind resources available (most residential wind turbine manufacturers know this, and some have designed turbines that can operate in wind speeds as low as nine miles per hour). The second problem is that unless your home has extremely low electricity demand, a wind turbine is probably not going to meet all of your needs, especially when the wind isn't blowing.

Your choice of turbine location is limited in most backyard wind settings. It's important to avoid obstructions such as trees (or even your house) that might cause wind turbulence that can seriously affect turbine performance. Installing a tall tower is one of the best strategies to deal with the turbulence problem (an eighty-foot tower is generally considered the minimum; even taller if you have trees nearby). However, tall towers sometimes run into problems with local zoning laws that restrict the height of structures, which is why backyard wind is mostly found in rural areas with less restrictive zoning.

Wind power can be used in either off-grid or grid-tied settings. Off-grid turbines normally have banks of batteries to store the (DC) electricity for windless periods, while grid-connected turbines basically use the grid as their "storage battery." Because there are rapidly spinning blades and other moving parts, routine maintenance of wind turbines is essential. New tilt-down towers make turbine maintenance a lot easier and safer than having to climb a fixed tower.

A residential wind project can cost anywhere from $16,000 to $80,000 depending on turbine and tower size and a wide range of other issues. This is another case where performing a home energy efficiency retrofit first makes a lot of sense. There are numerous utility, state, and federal incentives or tax credits for residential wind installations. Most municipalities require a building permit for a wind turbine (if they are allowed at all). States or utilities normally oversee the net-metering interconnection process. Depending on how close you live to them, neighbors might object to the potential noise or appearance of your turbine, so it's a good idea to check with them first before you go too far with a backyard wind project.

A wind/PV hybrid system offers a lot of advantages, since wind systems average higher output in the winter when winds are generally stronger while PV systems average higher output in the summer when the sun is higher in the sky. However, a combined wind/PV system can be very expensive.

Biomass Heat

As I mentioned in chapter 3, biomass has a lot of possible uses. But for the average family, biomass offers the most potential as a substitute for oil (or kerosene) or natural gas (or propane) to heat your home. The traditional approach has been to use firewood to fuel a wood-burning stove (or sometimes a boiler, or furnace), a strategy still popular in many (mainly rural) locations. More recently, wood pellets used to fuel specially designed pellet stoves, boilers, or furnaces have been gaining in popularity.

Because of their extremely efficient combustion process, pellet-fired heating appliances have very low emissions and often can be used in locations where regular woodstoves are restricted. The pellets (which look a lot like rabbit food) are generally made from waste sawdust or low-grade wood, and are normally packaged in forty-pound bags that are fairly easy to handle and store. Because of their relatively consistent size, wood pellets can be handled by automated systems, substantially reducing the amount of labor involved with pellet-fired appliances compared with a standard woodstove. Most pellet stoves only need to be fueled once a day—something that most woodstove users can only dream about. In addition, pellets are a lot cleaner and create less mess than firewood.

Pellets are generally sold by the ton (fifty 40-pound bags loaded on a pallet), which has the heating value of about one and a half cords of firewood and takes up about half the storage space of a cord of wood. Although the price of pellets has gone up and down in recent years along with supply and demand, it has been consistently less volatile than the price of oil and generally less expensive. Since your home heating appliance is probably your number one fossil fuel consumer, heating with renewable biomass can cut your carbon footprint dramatically, save you money, and increase your family's energy resilience.

The downside of pellet-fired appliances is that the fans and augers in them need electricity to operate, and the mechanical or electronic components can malfunction (although this is also true of most fossil-fueled heating appliances). Pellet stove prices range from around $1,500 to $3,500. Most woodstoves range in price from around $1,000 to $2,600. Before installing any biomass heating appliance, check with your town fire department, local building codes office, state energy office, or state environmental agency about wood-burning regulations that may apply to your location.

FIGURE 4-2. Pellet-fired boiler and stacked 40-pound bags of pellets in the author's basement. Photo by Greg Pahl.

Geoexchange

As I mentioned in chapter 1, ground-source heat pumps (geoexchange) can be used to heat or cool buildings using latent ground heat. This strategy can also be used to heat or cool your home. But unlike most other heating appliances, heat pumps are not based on combustion. Instead, heat pumps move heat from one location to another with extreme efficiency. In the winter, heat is extracted from well water (or sometimes from fluid in pipes buried in the ground) and then increased with the help of a compressor. The heat is then distributed through your home, usually by a series of air ducts. In the summer, the process is reversed and the heat from your home is removed and dissipated into the ground.

In milder climates, a similar technology, known as air-source heat pumps, can do the same thing, using latent heat in the air instead of the ground. Air-source heat pumps are considerably simpler and less expensive than ground-source geoexchange systems. Heat pumps will work in almost any part of the country although there may be some site-specific limitations, especially for ground-source heat pumps.

Geoexchange systems tend to use more electricity than a traditional fossil-fueled heating appliance, and are more expensive to install (but there is no fuel cost). Prices for geoexchange systems range from around $7,500 to $20,000 or more, depending on a wide range of variables. However, over the long term, geoexchange is the most energy-efficient, environmentally clean, and cost-effective space-conditioning system available, according to the Environmental Protection Agency (for more information on geoexchange, biomass home heat, and solar hot water, see *Natural Home Heating: The Complete Guide to Renewable Energy Options*, Greg Pahl, Chelsea Green, 2003).

Micro-Hydro

There is one local renewable energy resource that has the potential to provide all the electricity your household needs on a continual basis: micro-hydro. However, you need to know that the emphasis is on "potential," and that only a relatively small number of homesites might have the right combination of factors available. In addition, the approval process can be daunting. Nevertheless, if you have easy access to a river or stream, micro-hydro is attractive enough to make it worthwhile to investigate.

Micro-hydro refers to small water-powered systems that generate one hundred kilowatts or less of electricity, and in the case of an individual household, a lot less. A ten-kilowatt installation can easily provide enough electricity for a large home or small farm, and even a tiny five-hundred-watt generator can power most of the appliances in a very energy-efficient home. An extremely small generator can accomplish this seemingly impossible task because it's generating twenty-four hours a day, 365 days a year. The excess DC electricity (from times when there is no demand) is stored in a bank of batteries for later use.

Because they are so small, micro-hydro systems use a diversion strategy known as run-of-river that normally does not involve a dam or water storage reservoir. The small quantity of water required generally flows through a pipe down to the generator and then is discharged back into the river (or stream); there is virtually no negative environmental impact. A micro-hydro installation can be off-grid (a DC system) or grid-connected (an AC system). For a number of technical reasons, AC systems require larger generators (in the two- to three-kilowatt range) and more water to run them.

Because most small micro-hydro systems generate direct current, they can be combined with wind or PV systems to create an all-season hybrid. Micro-hydro systems generally cost about $10,000 to $20,000. There are no federal tax incentives for residential micro-hydro systems, but there are some incentives in a few states (check the U.S. Department of Energy's DSIRE database at www.dsireusa.org). The regulatory and approval process for micro-hydro, which in many cases can involve federal, state, and local regulators, can be lengthy, expensive, and frustrating.

Urban Solutions

If you own a town house or condominium, your individual household's options to generate your own electricity are more limited, but not impossible. Sarah Lozanova and Kirl Lozanov live in a Chicago condominium, and managed to convince their condo owners' association to allow them to install a 1.7-kilowatt PV system on the condo's roof. What's more, they also managed to get their friends and family to contribute toward purchasing the PV system as their main wedding gift!

As a first step in planning their PV system, the couple reduced the amount of electricity they used by installing an energy-efficient refrigerator, washer, and dryer, as well as compact fluorescent and LED lightbulbs. And since they line-dry their clothes, the dryer is rarely used. A home energy audit also showed that their VCR, DVD player, and stereo were "leaking" electricity even when turned off; these appliances are now plugged into switched power strips.

Part of the agreement with the condo owners' association stipulated that there would not be any roof membrane penetrations, so the PV array was mounted on an adjustable ballasted pan system weighed down by cement blocks. The PV panels can be moved if maintenance work on the roof is needed. The system generates virtually all of the couple's electricity, and the remainder is fed into the grid under a net-metering arrangement.

"Although I believe solar technology will mature significantly in the decades to come, I'm excited to be able to tap into this vast energy source now," Sarah Lozanova says. "I like coming home to see how much energy the system generated throughout the day and knowing that I'm part of the renewable energy solution."[8]

Not all condos have an appropriate available roof space—or such a cooperative homeowners' association board. Condos generally face a number of challenges to going solar, but especially financing that involves multiple owners, condo regulations, and other complex legal issues. But even if you can't work out an agreement to install a large PV array on your building, you might be able to install a small PV-powered lighting system for one room. A little solar-powered light is better than none.

Another exciting possibility is the increasingly popular solar gardens movement. This relatively recent phenomenon offers renters, condo owners, and other community members the opportunity to own solar panels in a common array located on buildings or sites that have good solar exposure. A wide range of different legal and financial models are being used by the groups involved in this movement, some of them relatively untested. Nevertheless, this is one of the best ways to give people who otherwise would not be able to generate their own electricity the chance to do so (for more on solar gardens, see chapter 6).

PART THREE

COMMUNITY ENERGY RESILIENCE

5

Community Energy

You got to be careful if you don't know where you're going
because you might not get there.
—Yogi Berra

In recent years there's been a resurgence of interest in community energy, the idea that most if not all of the energy consumed in a community should come from—and be owned and controlled by—the community itself. Community energy initiatives are now emerging across the country, and they take a variety of forms. Community-supported energy (CSE), for example, is similar to community-supported agriculture (CSA), except that instead of investing in greater food resilience, local residents invest in greater energy security and a cleaner environment. Another model is community-owned renewable energy (CORE), which emphasizes the key role of local ownership. Whatever form it takes, community energy encourages some new thinking about our relationship with energy. In many cases it also involves neighbors working with one another for the well-being of the community as a whole—an old idea that in some parts of the country goes back to colonial and even pre-colonial times. In any case, community energy admittedly involves a lot of hard work by community members, but this is part of the process of taking greater responsibility for our own energy consumption and production.

Why Community Energy?

The main goal of corporations is to generate profits, increase "shareholder value," and in most cases provide handsome pay and benefits for top management. This is certainly true of most companies in the energy sector. For example, Lee Raymond, ExxonMobil's former chairman and CEO, received a $400 million retirement package in early 2006. In 2008, the company went on to make record profits of $45.2 *billion* while most Americans struggled to pay for gasoline at the pump.

The main goal of most local communities, in contrast, is to provide for the general welfare of the people who live in them; they spend what limited resources they have at their disposal to offer needed public services. The contrast is stark: profits versus public good.

This is why community energy undertakings make so much sense. They aim not to make a few directors and investors rich, but rather to provide energy security for local residents and businesses and long-term stability in local energy costs, all while providing local jobs and keeping energy dollars circulating in the local economy. High profits and big management benefits packages are not normally part of the picture. This offers local communities some flexibility in planning for projects that might not pass Wall Street's test for high returns, but do pass Main Street's test for community benefit.

In chapter 3, we saw that many local renewable energy resources—solar, wind, hydro, biomass, biogas, liquid biofuels, and geothermal—are potentially available to be harnessed by local projects. Not all of these resources are available to all communities, but nearly every community will have at least a few of them. We also saw that many of these resources can be used to generate electricity. A growing number of experts see electricity as the foundation for a completely revised transportation system.[1] The fact that there are multiple opportunities for generating local electricity offers the possibility that at least some of this transport can be locally powered. For urban areas this normally involves electrified public transport, while rural areas would tend to rely more on plug-in hybrid electric vehicles.

The multiple local benefits of community energy initiatives are evident, especially as compared with conventional energy undertakings (see the "Core Principles of Community Energy" sidebar). Obviously, community energy can't fully address the near-total dependence of the national economy on fossil fuels and its massive existing infrastructure; that issue will take decades—and billions of dollars—to resolve. But community energy can at least offer local economies some protection from future disruptions caused by energy supply restrictions or price volatility—or both.

Core Principles of Community Energy

Community energy initiatives are not simply smaller versions of the utilities and multinational energy corporations with which we're familiar. They are fundamentally different. Below are the four key principles I believe should guide all community energy initiatives.

1. COMMUNITY OWNERSHIP, COMMUNITY BENEFIT
Conventional energy businesses must put profit first. Most community energy projects, however, take a deeper view—ensuring that they meet the broader needs of the community (including the health of the local economy and environment). Local ownership is key to this difference. It not only shifts the results of decision makers' actions into their own backyards, but also reinforces the accountability of those decision makers to the people they're serving.

2. RENEWABLE, LOCAL, AND DISTRIBUTED
Renewables by definition won't run out, so they're ideal for building local energy security. Renewables are generally available across a wide geographic area, making it difficult for a private entity to monopolize the sole point of production, and making the energy system more resilient as a whole thanks to its distributed nature. Finally, because renewables are much lighter on the environment than nonrenewables, they're ideal for passing on a healthy, clean community for future generations.

3. ADAPTIVE RESILIENCE
Unlike most private businesses, which can move, merge, or even dissolve when needed, the community is not going anywhere—so the ability to adapt to changing conditions over decades, even centuries, is essential. If done right, community energy initiatives are less vulnerable to external shocks—whether these be price spikes or breaks in supply chains—and are more able to adapt to changing conditions.

4. CONSERVATION FIRST

With the end of cheap and abundant fossil fuels, we have little choice but to reduce the overall amount of energy we consume. We've grown our modern economy and built our communities on inherited fossil fuels. Transitioning to living off the more modest stream of renewable energy will require some cutting back of excess—but as most of Europe knows, a very high quality of life is thoroughly possible and enjoyable with only modest energy consumption.

Plan Ahead

Most communities have some sort of municipal plan, and many use it to engage local residents in a discussion of their shared values and concerns to prepare for the community's future. Developing a community energy plan is a similar exercise, but with a focus on energy. Thinking about community energy normally involves taking a close look at present energy use, and then developing a plan to reduce it and shift to as many local renewable alternatives as possible. A comprehensive community energy plan that reduces consumption will almost certainly contain increased energy efficiency and conservation strategies for municipal facilities and often for the entire community as well.

Community energy planning can also be integrated with land-use and transportation plans to promote more efficient and less energy-intensive travel opportunities like walking, biking, and public transit. In many locations energy issues transcend municipal boundaries, so community energy planning might also include a regional element.

Public engagement is crucial. If the planning process for a renewable energy project involves local input from the beginning, it becomes a community project in the minds of the participants—especially if the project is locally owned. This automatically eliminates most of the potential opposition to the project because they see that it benefits the community rather than some faceless out-of-state corporation. And because the project was planned by local residents and community leaders, any objections or concerns are generally resolved as part of the local planning process. For example, the town of Alderson, West Virginia, recently embarked on a comprehensive Community Energy Plan (CEP) with a grant from the Appalachian Regional Commission to develop a community-based approach to energy and economic planning. The project has the strong support of the town council, which approved a resolution supporting a "green team" of local citizens to help implement and guide it.[2] Local ownership and control helps the community create a project that meets its particular needs while addressing issues of size, scale, and location—issues that local stakeholders are best positioned to tackle.

Another major benefit of community energy is that the collaborative effort required to get a renewable energy project up and running tends to strengthen the community in many ways. As community members work together on an exciting (and often challenging) initiative, they develop skills and relationships that provide the foundation for additional collaborations. Local knowledge and expertise in renewables increases, strengthening the community's energy resilience. One successful project can lead to another—and then another.

Getting Started

So how exactly do you set up a local community energy project? That's not an easy question to answer because there are many different possible financial and ownership strategies, and selecting the right one depends on a wide range of local issues. Some community energy initiatives focus on energy efficiency, while others may promote a particular technology such as solar hot water. Others may involve a larger-scale community energy project such as a community wind farm or group net-metered PV installation. Some communities might combine these elements. There is no one "right way," just the way that works best for your community.

However, if your community wants to pursue a collaborative energy project such as a large-scale wind turbine, solar farm, small hydro, or perhaps a neighborhood biomass district heating system, you will need to consider the options that are appropriate to the particular technology. A good wind turbine site will be quite different from a good small hydro site, and the regulatory and approval processes that apply are going to differ as well. A carpenter has a wide range of tools to choose from to get the job done. A community energy project developer has a large number of tools to work with as well. You and your community will have to pick the right combination for your particular location and type of project.

In this section, we'll review in some detail your options for the two most important foundations of your project: financing and legal structure.

Financing Strategies

One of the biggest issues for most projects is how to finance them. Renewable energy technology tends to be fairly expensive and usually requires a lot of upfront equity and debt financing as well as seed capital to cover early development costs. This can be a daunting challenge for most communities or local organizations, depending upon who is actually organizing the project. Nevertheless, there are a number of financial tools available. You will probably want to take advantage of as many as possible. For additional ideas on local financial strategies, see the first book in this Community Resilience Guides series, *Local Dollars, Local Sense* by Michael Shuman.

TRADITIONAL FUNDING

One of the first places to look for financing for small, locally owned projects is a local bank. Local bankers have an obvious interest in their community and in establishing or maintaining financial relationships with local residents. They may not be familiar with the latest renewable energy technologies, but if the benefits to the local community are carefully explained, some bankers may be persuadable. The Acorn Energy Solar One, LLC solar PV project in Vermont (described later in this chapter) received some of its financing from a local bank. If there is no local bank in your community, a regional bank is the next possible source. There are also a number of commercial finance companies that specialize in renewable or green energy projects. Vermont's Community Energy Exchange looks like a promising new model (see www.communityenergyexchange.com) for matching small investors with small community energy projects using an online platform (see "Resources" at the end of the book and the Burlington Cohousing Solar Project in chapter 6).

It is also possible that a local business or cooperative might be willing to be an equity investor or participant in a local renewable energy project, depending on their appetite for federal or state tax credits. Some states also offer low-interest loan programs for renewable energy projects. Admittedly, obtaining a loan of any kind has been a challenge for most small businesses since 2008. But if your project is well conceived and structured, and if you are able to take advantage of some of the incentives or tax credits that are often available, you should be able to arrange financing.

PROPERTY ASSESSED CLEAN ENERGY PROGRAMS (PACE)
Another potential tool is a local Property Assessed Clean Energy program. These programs enable local governments to finance renewable energy and energy-efficiency projects on residential, commercial, and industrial properties, eliminating the main barrier to these projects: the large upfront cost. The local government does this by creating an improvement district and issuing a bond secured by real property in the district; the bond proceeds are then available for funding renewable energy or energy-efficiency projects. The property owners who take advantage of this funding then repay the debt service on the bond in fixed payments as part of their property tax bills. If the property is sold, the debt stays with the property and is assumed by the new owner, who will also benefit from the improvement. Participation is completely voluntary, and only those who choose to participate in the program pay the costs of the additional assessment.[3]

PACE programs have begun to spread across the nation (twenty-seven states plus the District of Columbia), although they were suspended in July 2010 due to regulatory issues with the Federal Housing Finance Agency and mortgage giants Fannie Mae and Freddie Mac. Some states, however, have since enacted recent PACE legislation enabling the commercial portions of the program to continue unaffected, and there is now an active bipartisan national effort to allow the programs to continue (see PACEnow.org). While PACE programs may not be useful for collaborative community-scale projects, they can be extremely helpful for private residential and commercial projects in the community.

PRODUCTION TAX CREDITS (PTC)
Another possible tool is the federal Renewable Electricity Production Tax Credit. The PTC reduces the federal income taxes of qualified taxpaying owners of renewable energy projects based on the electrical output (measured

in kilowatt-hours). Qualified resources include landfill gas, wind, biomass, hydroelectric, geothermal electric, municipal solid waste, anaerobic digestion, tidal power, wave energy, and ocean thermal. The PTC is generally available for the first ten years of operation and provides 2.2 cents per kWh for wind, geothermal, and closed-loop biomass, and 1.1 cents per kWh for open-loop biomass and other eligible technologies.[4] The credit has experienced numerous congressional cycles of expiration, renewal, and revision, making it difficult for the renewable energy sector in general (and businesses representing the specific technologies in particular) to make long-term financial plans.[5] According to the program deadlines as of this writing, the project must be placed in service by January 1, 2013 (for wind) or January 1, 2014 (all other technologies).

The PTC is mainly meant for corporations with a large tax credit appetite, and is difficult for smaller groups such as farmers, schools, municipal utilities, and cooperatives to use due to technicalities in federal tax regulations. Nevertheless, many community energy projects (especially wind power projects in the Midwest) have found creative ways to make use of the tax credit.[6] Two of the most famous examples are known as the Minnesota Flip and the Wisconsin Flip, both originally developed for community wind projects in those states. The idea is that local investors form a limited liability company (LLC) to do all of the preliminary work of organizing the project, and then market the project to a tax-motivated equity investor who can take advantage of the PTC and accelerated depreciation for an initial ten-year period. Typically, the equity investor owns around 90 percent (or more) of the project for the first ten years. Then, when the PTC has expired, the ownership "flips" to the local LLC. This strategy can sometimes be adapted to other types of renewable energy projects. Of course, none of these convoluted strategies would be necessary if the United States had a coherent national energy policy, including a national feed-in tariff (more on this later). For additional information on PTCs, see the DSIRE database (www.dsireusa.org).

BUSINESS ENERGY INVESTMENT TAX CREDITS (ITC)
Yet another federal corporate tax credit that supports renewable energy projects is the Business Energy Investment Tax Credit. The ITC reduces federal income taxes for qualified tax-paying owners of renewable energy projects, and is earned when the equipment is placed in service. The eligible technologies

include solar hot water, solar space heating, solar photovoltaics, wind, biomass, geothermal electric, fuel cells, geothermal heat pumps, combined heat and power (CHP) co-generation, and others. The amount of the tax credit generally ranges from 10 percent for geothermal, micro-turbines, and CHP, to 30 percent for solar, fuel cells, and small wind. There are no limits on the maximum incentive on small wind turbines placed in service after December 31, 2008, and no limits on all other eligible technologies regardless of their installation dates. The eligible system size varies, depending on the technology.[7] In order to qualify for the ITC, the system owner must be a taxpaying entity and the system must be placed in service before December 31, 2016.

In addition, all ITC-eligible technologies, as well as large wind projects, can take advantage of Modified Accelerated Capital-Recovery System (MACRS) accelerated depreciation allowed by the IRS (see www.irs.gov/publications/p946/index.html). A five-year depreciation schedule is generally allowed, although certain biomass projects can use a seven-year depreciation schedule. In addition, bonus depreciation that allows taxpayers to deduct 50 percent of the value of eligible systems in the first year currently extends through 2012.

New Markets Tax Credits (NMTCs)

Established in 2000, New Markets Tax Credits were intended to spur new or increased investments located in low-income communities. They have been used to help finance solar projects as well as—in at least one community—a wind project, built in 2010 in Washington State (Coastal Energy Project LLC). NMTCs provide an investment tax credit of 39 percent over seven years for a Qualified Equity Investment (QEI) in a Community Development Entity (CDE). The CDE, in turn, channels "substantially all" of the QEI into a loan or equity investment for a qualifying low-income business. One particularly attractive feature of this strategy is that it appears to be possible to make use of an NMTC in addition to a PTC or ITC. While this is not a tool for everyone, it's worth investigating.[8]

Clean Renewable Energy Bonds (CREBs)

CREBs are a federal loan program for municipalities (including school districts), municipal utilities, and rural electric cooperatives to finance renewable energy projects. The same list of qualifying technologies used for PTCs generally applies to CREBs. CREBs are issued with a 0 percent interest rate; the borrower only pays back the principal of the bond while the bondholder receives a federal tax credit. The federal Energy Policy Act of 2005 established CREBs to finance public-sector renewable energy projects. Since then, the guidelines have been amended several times. Participation in the program is limited by the volume of bonds allocated by Congress (and announced by the IRS), and those allocations have varied. CREBs are part of the financing package being used for Hamilton, Ohio's Meldahl Hydropower Project (see chapter 8).

RENEWABLE ENERGY PRODUCTION INCENTIVE (REPI)
This is a federal performance-based incentive for projects owned by local municipal governments (including school districts), state governments, municipal utilities, rural electric cooperatives, and native corporations. Eligible technologies are similar to those for PTCs, plus solar thermal electric and photovoltaics. The payment amounts to 2.2 cents per kWh (subject to availability of annual appropriations). The production payment only applies to electricity sold to another entity, and actual incentives have generally been lower in recent years due to lack of full funding. REPI was designed to complement the federal PTC, which is available to businesses that pay federal corporate taxes.

USDA RURAL UTILITY SERVICE (RUS)
Section 6108 of the 2008 Farm Bill expands the U.S. Department of Agriculture's authority to loan to renewable energy generation projects, even if those projects are not serving traditional rural markets. This clause potentially makes low-cost debt financing from USDA's Rural Utility Service available to a wide range of renewable energy projects, even those not associated with rural electric cooperatives. A RUS loan was used for part of the financing of the Fox Islands Wind Project in Maine in 2009. This was the first time a RUS loan was made to a wind project on a project finance basis (more on Fox Islands Wind in chapter 7).[9]

STATE AND UTILITY INCENTIVES
In addition to the federal programs just listed, a dizzying array of state, utility, and local incentives, rebates, and policies can be tapped to assist many

community renewable energy and efficiency projects. Visit the DSIRE website (www.dsireusa.org) and click on your state to see what is available. It's important to understand that the presence or absence of local or state incentives or other policies can dictate which of the previous financial and ownership strategies are viable in your location. Check with financial and legal experts in your region.

Legal Options

Because community-supported energy is a relatively new idea in the United States, the legal and other structures to support it are still, to some extent, works in progress. Nevertheless, there are a number of traditional (and a few not-so-traditional) ways of approaching ownership and development models for local energy projects. For most community renewable energy projects, some form of joint ownership makes sense.

Existing Local Public Entity

In some cases, depending on local circumstances, the simplest approach might be to go with an existing local public entity rather than take the time and expense to create a new one. Towns and school districts in most locations can develop renewable energy initiatives, and this is the ownership strategy many communities follow with their first project(s). The Ellensburg Community Solar Project in Washington (chapter 6) and the Willmar Community Wind Project (chapter 7) are examples of this strategy. For wind, solar PV, hydro, and other projects that generate electricity, any excess power may be sold to the grid through net metering. Financing for municipal projects might be accomplished with CREBs. Many entities eligible for CREBs might also qualify for REPI, although it's not clear whether both programs could be used for the same project.

A relatively new legal option available in six states, Community Choice Aggregation (CCA), provides an alternative and in some ways simpler route for utilizing public entities to exert more control over the power the community consumes—but without actually owning the plant and transmission infrastructure. Through CCA, a group of local governments can aggregate the demand of their respective communities (the residential and business ratepayers, as well as government institutions) and then select and purchase energy at superior terms on their behalf. "The reasons to pursue CCA vary by community," says Shawn Marshall of the Marin Energy Authority, California's first CCA program, "but chief among them are lower electricity costs, cleaner energy supply, greenhouse gas reduction benefits, and the development of local generation assets."[10] CCA is now available in California, Illinois, Ohio, Massachusetts, Rhode Island, and New Jersey (see www.leanenergyus.org).

PARTNERSHIPS

A general or limited partnership is a relatively simple and flexible structure in which the profits and liabilities from a project are split among the partners. It's easy to set up, and almost any requirements (within reason) can be included. It has the disadvantage of holding the partners personally liable for debts incurred by the partnership, although insurance can be bought to limit the actual liability. The large upfront costs for a project are generally borne by the partners, which limits participation to people with considerable financial resources.

LLCs

A limited liability company (LLC) is a corporate entity that can be the owner of a renewable energy project. An LLC offers many benefits: flexibility in ownership, management, and decision-making power; enhanced liability protection; and profit distributions. Individual investors normally buy shares in the LLC. Gains and losses are allocated to the shareholders for tax purposes. One of the main advantages of LLCs is that they offer legal protection to the owner/ investors in the event of financial problems or litigation. A financing entity (if used) can receive tax benefits from funding the project (in this case, for tax purposes the LLC would be described as a "pass through entity" that offers the tax advantages to the financing entity). There are no federal corporate taxes, but profit distribution passes through to the investors and is treated and taxed at the rate relevant to those investors. Federal production tax credits may be available. A key disadvantage is that the LLC by itself (without a financing entity) probably won't have sufficient "tax appetite" to take advantage of PTCs.

Under certain circumstances a low-profit limited liability corporation (L3C) might be a useful organizational structure. The L3C differs from an LLC in that its main purpose is to achieve a social benefit rather than a profit—something that might well describe a community renewable energy project (L3Cs are allowed to make a profit, it's just not the main stated goal). An L3C must "significantly further the accomplishment of one or more charitable or educational purposes," and would not have been formed except to accomplish those purposes. L3Cs offer all the benefits of an LLC plus some potential advantages in possibly attracting financing from private foundations for program-related investments (investments that help the foundation achieve one or more of its main goals).[11] This is a relatively new structure, not yet available in all states, but it might be worth looking at.

Co-ops

A cooperative is a business owned by its customers or workers or both. In this model, local investors pool their resources and use their capital to leverage debt financing. Cooperatives today follow the model provided by the Rochdale Principles first developed in Rochdale, England in 1844. These seven principles are voluntary and open membership; democratic member control; member economic participation; autonomy and independence; education, training, and information; cooperation among cooperatives; and concern for community.

Co-ops offer a number of advantages. There are three core principles of co-ops that make them potentially useful for community energy initiatives: user benefit, user control, and user ownership. The details vary a bit depending on the type of business a co-op is engaged in, but an obvious user benefit for an energy co-op could be lower or more stable electricity prices or fuel costs. Control remains with the members, who elect a board of directors from the membership. User ownership in a co-op is usually accomplished by purchasing one share of voting stock and doing business with the co-op.[12] In the case of a co-op formed for a renewable energy project, the members come together specifically to initiate the project and buy or use the energy or fuel produced by the project. The disadvantages are that cooperatives normally cannot take advantage of PTCs or CREBs due to lack of "tax appetite." Raising large amounts of capital or securing member loans can be problematic with many co-ops since members may not be able to raise sufficient capital themselves for a large start-up venture.

In today's highly competitive, corporate-dominated society, a small, local co-op is a lot of work and not an easy model to follow, but it involves the kind of collaborative thinking that will be required in the new, localized, more self-sufficient economy of the twenty-first century.

MULTILATERAL LICENSING AGREEMENT

A multilateral licensing agreement is another new ownership model that might be useful for some community-owned renewable energy (CORE) group net-metering projects (more on that in a moment) using solar PV.

A multilateral licensing agreement offers a lot of flexibility: It can define the relationships among the landowner, the members, and the facility's management and operation. It can be structured to benefit single or multiple tax-advantaged investors who purchase the solar PV system and receive the investment credit with limited paperwork and headache from the IRS. A simple contract defines how members want their relationship to be governed, and covers who can become members, who administers the group, who the landowner for the project site will be, changes in membership, and so on.[13]

A multilateral licensing agreement was suggested in a 2010 study, *Vermont Group Net Metering Information & Guidelines*, produced by the Vermont Group Net Metering Team for the Clean Energy Development Fund and Powersmith Farm (a 250-kilowatt solar photovoltaic model was used for the study). It's easier to set up than a cooperative or LLC and involves a lot less paperwork. Because this model is so new for community energy projects, it is essentially untested; still, it's worth consideration due to its simple structure and relative low cost.

PUBLIC UTILITY DISTRICT (PUD)

Also known as a People's Utility District, a PUD is a community-owned, locally regulated utility created by a vote of local citizens that provides utilities such as water, electricity, sewer service, and (more recently) telecommunications to people who live within the district. PUDs are created by cities, counties, or other local governmental entities. Most of these districts were formed in the 1930s or '40s and are nonprofits. Many of the PUDs in the Pacific Northwest were created specifically to take over the territories then served by investor-owned utilities, and were part of a larger movement, supported by the Grange and others, that encouraged public ownership of electric utilities. There are twenty-eight PUDs in Washington State; twenty-three provide electricity, nineteen provide water and wastewater services, and thirteen offer wholesale broadband telecommunications.[14] There are eight PUDs operating in Oregon, and dozens of municipal utility districts, electric cooperatives, and other public utilities in surrounding states.

Net Metering and Feed-in Tariffs

While community energy projects can choose their own financial and legal structures, they operate within the larger context of utility, state, and federal policy. Two of the most important policy developments for community energy in recent years are group net metering and feed-in tariffs.

GROUP NET METERING

As mentioned in chapter 4, net metering allows owners of small, grid-connected electricity generating systems (typically wind, solar, or hydro) to receive a credit on their electric bill for the electricity produced above what they consume in their home or business. Group net metering takes this concept one step farther and allows a group of customers served by the same utility to combine their meters as a single billing entity to offset that billing against electricity production from a large net-metered system. This is sometimes referred to as meter aggregation or virtual net metering. In most cases, the net-metered electricity generating system can be located almost anywhere in the same utility territory serving the group members, offering a lot of flexibility for finding a good site. In addition to homeowners, group net metering offers residents of multifamily homes and apartment dwellers the possibility of participating directly in renewable energy projects.

Vermont, Maine, Oregon, Washington, and West Virginia offer group net metering. California, Colorado, Massachusetts, Nevada, New York, Pennsylvania, Rhode Island, and Utah offer limited group net metering in certain circumstances. More states are sure to follow this trend.

FEED-IN TARIFFS

In the United States, large corporate interests have tended to dominate not only the traditional energy sector but also the renewable energy scene, and government policy has generally reflected that centralized model for decades. But this is changing. The agent for this change is the renewable energy feed-in tariff (FIT). A feed-in tariff offers the promise of a different model in which almost anyone can become a renewable energy investor and owner.

A feed-in tariff is a policy mechanism designed to encourage investment in renewable energy technologies that generate electricity. This is accomplished by guaranteeing grid access and offering long-term contracts to renewable energy producers, including homeowners, business owners, farmers, and private investors. The prices offered in the contracts normally vary depending on the cost of generation for the different technologies, and often decrease over time to encourage technological innovations. One key difference between a FIT and net metering is that a FIT pays for *total* generation, whereas net metering only pays for *excess* generation.

The first type of feed-in tariff was implemented in the United States in 1978 with the National Energy Act, which contained the Public Utility Regulatory Policies Act (PURPA). There was a provision within PURPA that required utilities to purchase electricity generated by qualified independent power producers at attractive (to the producers) rates. This requirement made PURPA highly unpopular with most U.S. utilities, which resisted it for years.

However, feed-in tariffs (also known as standard offer contracts or advanced renewable tariffs) were particularly effective in Germany, which implemented its first *Stromeinspeisungsgesetz*, or electricity feed-in law (EFL), in 1991. Since revised a number of times, the EFL set a stable foundation for wind power and propelled Germany to the forefront of installed wind capacity for many years. Following Germany's successful lead, many other European nations, particularly Spain and France, adopted similar feed-in tariffs that have more recently encouraged the growth of solar power as well as wind.

As of 2011 feed-in tariffs have been enacted in over fifty countries, but those in Germany, France, and Spain are considered the best. The feed-in tariffs in those three countries are called advanced renewable tariffs to distinguish them from earlier, simpler programs. Unlike PURPA in the United States, which was based on what turned out to be inaccurate projections, these European FITs are based on actual renewable energy project economics (plus a reasonable return on investment). This is why these FITs have been so effective.

ONTARIO STANDARD OFFER CONTRACT

Meanwhile, back in North America, the lack of comprehensive national feed-in laws continued to hamper renewable energy projects in general, and community energy projects in particular. This changed dramatically in the spring of 2006, when the Ontario Standard Offer Contract (SOC) was announced by provincial officials. Patterned after European feed-in tariffs, the Ontario SOC allowed homeowners, landowners, farmers, cooperatives, schools, First Nations, municipalities, and others to install renewable energy projects up to ten megawatts in size and to sell the power to the grid for a fixed price for twenty years. Only four technologies were recognized, however, and the tariff was only split into two categories, one for solar PV, and another for all other technologies.

The Ontario Sustainable Energy Association (OSEA), an Ontario community-based renewable energy group, was instrumental in putting the Standard Offer concept on the political agenda in the province. They launched a campaign in early 2004 to adapt the European policy to Ontario so OSEA's members could form cooperatives to install wind turbines, solar panels, biogas digesters, and small hydro projects. "This type of broad local ownership structure is what we at OSEA call Community Power," says Deborah Doncaster, former executive director of OSEA. "A very important aspect of Community Power is that it is locally owned and developed."[15]

However, due to technical limitations within the original program, the Ontario SOC only resulted in about 150 megawatts of projects by mid-2009. The program was substantially revised in 2009, with the introduction of the Ontario Green Energy Act, the most progressive renewable energy policy in North America in decades. A revised and enhanced feed-in tariff (the Advanced Renewable Tariff) was a key feature of the new program. As a result, in April 2010 the Ontario government announced a total of $8 billion in new renewable energy deals—with 184 wind, solar, hydro, and landfill gas projects receiving long-term contracts totaling around twenty-five hundred MW of green power.[16] This included eighty MW of contracts to homeowners for solar PV, and almost one-fifth of all contracts were granted to homeowners, farmers, and community and aboriginal groups.[17]

U.S. FEED-IN TARIFFS

Following Ontario's lead, more than a dozen provinces, states, and other jurisdictions in Canada and the United States are now using or considering some form of feed-in tariff. Since 2008, the states of California, Vermont, Maine, Washington, Wisconsin, Oregon, Indiana, and Hawaii have enacted feed-in tariffs with differing provisions and program limitations. In most cases they tend to be timid and restrictive (thanks largely to ongoing resistance from utilities) but generally qualify as reasonably good first attempts (it took Germany more than ten years to get its feed-in tariff right). There are also a number of munici-pal programs in Gainesville, Florida (see chapter 13), San Antonio, Texas, and Sacramento, California. More states and municipalities are expected to follow.

To further complicate an already confusing scene, in addition to the state FITs there are a number of utility-based "FITs" as well. However, unlike the successful advanced renewable tariffs, these utility programs are not cost-based, and not really intended to stimulate large amounts of renewable energy. Opponents of FITs (many utilities among them) tend to complain about the "foreign" roots of feed-in tariffs, but ironically U.S. utilities receive the same "cost-recovery plus profit" treatment for their own conventional projects from U.S. regulators.[18] Whether the federal government ever plays a productive role in developing a national advanced renewable feed-in tariff is an open question.

Barriers to Local Energy

As we have seen, renewable energy technologies can easily be scaled up or down to meet community needs, so there are really no technical barriers. There are, however, quite a few other problems that have stymied many local energy initiatives for years. Right at the top of the list is a lack of a comprehensive and effective national energy policy in the United States. For four decades, U.S. administrations have repeatedly failed to reduce our petroleum consumption or imports. Even a simple measure like a carbon tax—which would tax the carbon content of fossil fuels and send price signals to spur carbon-reducing investment—remains stalled while many inside the Beltway continue to pander to the coal, oil, and natural gas industries, as well as the big utilities.

The counterproductive role that utilities have historically played cannot be overstated. The early local wind power initiatives in Europe were resisted vigorously by utilities and regulatory agencies. That same pattern repeated itself in North America, where most utilities viewed local renewable energy proposals as challenges to their monopolies. Roadblocks were set up against making affordable or convenient connections, tariffs were set too low to allow for project financing, and caps were set too low for meaningful net-metered projects. In recent years, some utilities have become more supportive of local renewables, and a few have actually become proactive after they ran the numbers and realized that small-scale distributed generation saved them money during peak demand times by reducing the need to buy expensive power on the spot market. Nevertheless, since the highest profit for many utilities often comes from long-distance transmission of electricity, these utilities tend to support large-scale renewable generation projects far from the ultimate end users.

Many federal, state, and local regulatory agencies have also been significant barriers for community energy projects. The people who staff these agencies are not necessarily opposed to renewable energy; they simply are applying inflexible rules and regulations that have evolved over many decades of reliance on large-scale, centralized, corporate-dominated energy in this country. Bureaucracies tend to resist change, and it usually takes years of strong political pressure to enact reforms. This has certainly been the case with many of the rules and regulations relating to community renewables. In some cases, new, more flexible rules are needed, while in others there need to be exemptions for small-scale projects that have broad community support.

Small hydro is the most glaring example of this scenario. The same federal and state regulations that would apply to constructing a new Hoover Dam generally apply to even the smallest community hydroelectric project. The cost of the painfully slow and complex permitting process involving the Federal Energy Regulatory Commission (FERC), the U.S. Army Corps of Engineers, and others, can easily exceed the costs of the actual project. There is really no excuse for this kind of foolishness, but it has taken many years for FERC to finally respond to criticism from local small hydro supporters. Finally, in August 2010, FERC unveiled its Small/Low-Impact Hydropower Program website[19] intended to help small hydro developers win quicker approval of their projects (see Logan, Utah, in chapter 8). This represents the elimination of one of the most seemingly intractable regulatory hurdles in the community renewables sector and has already encouraged a growing number of small hydro project permit applications in the past few years. Some people within the hydro industry complain that the FERC process is still too slow, but there has definitely been some progress in any case.

Another barrier has been the constant change in incentives, tax credits, and other support for renewables over the years. Many of these supports have been part of federal legislation that often includes sunset provisions or annual renewal requirements. As the political winds inside the Beltway have shifted one way or the other, so have the supports. This has made it extremely difficult for various renewable industries to make any long-term plans, and made it virtually impossible for smaller community-scale projects to obtain financing. However, with the proliferation of net metering, group net metering, and feed-in tariffs, many of these barriers are beginning to crumble, and communities that pursue local energy projects are far more likely to succeed today than they might have been just a few years ago.

Not in My Backyard (NIMBY)

However, there is still one problem that has not gone away: the "not in my backyard" (NIMBY) response. This is not a simple issue with simple answers. One important aspect of community energy involves taking greater responsibility for our own energy consumption, rather than relying on utilities to do it for us *in someone else's* backyard. This inevitably requires some compromises.

While almost any energy project is going to have some environmental impact and at least a few opponents, some of the most vocal opposition to renewable energy has been centered on large-scale, corporate-dominated wind farms. These projects, with their huge, four-hundred-foot-plus-tall, industrial-scale turbines and flashing lights, can admittedly overwhelm local landscapes if they are not carefully sited. In the case of most large offshore wind farms, such as Cape Wind planned for Nantucket Sound, the turbines are to be located far enough offshore to eliminate most of the objections raised by the project's opponents.

In the case of large-scale wind projects located in small rural communities, the situation is less clear-cut. On the upside, wind farms rarely add any significant municipal expenses, and the host communities generally need the land lease payments, tax payments, or production royalties for their municipal budgets. On the downside, if the turbines are located too close to homes there is some possibility of noise and light pollution problems as well as shadow flicker caused by the spinning blades under certain circumstances. There can also be some negative environmental impacts, but most of these problems can be eliminated by careful siting and project design. There is no such thing as an entirely benign energy project, regardless of the technology; it's a question of trade-offs.

I have no problem with people who object to large-scale wind farms owned by out-of-state corporations that will be taking the profits (and most of the electricity) out of town, as long as these same people are willing to support smaller, community-scale wind or other renewable energy projects that are locally sited and owned. Community energy projects can eliminate most of the NIMBY problems by including the community in the planning from the very beginning and by demonstrating that the project directly benefits the community through local ownership. Community wind projects typically involve just a few turbines, as opposed to dozens or hundreds. It's a compromise that quite a few communities have been willing to make.

While other renewable energy technologies also have some opponents, community solar is one strategy that has few if any NIMBY problems associated with it, making solar especially attractive from this standpoint. In the end, it all comes down to a fairly simple issue: If we want to continue to have a reliable source of electricity in our communities, we need to be willing to accept more responsibility for generating some of it in our own backyards.

Models

Not many years ago, when I would give a public presentation about community energy, one of the first questions was, "What's that?" Not anymore. Rapidly growing numbers of local communities, groups, and organizations around the country (and around the world) are focused on building community energy resilience. Those efforts vary from one community to another, but the common thread running through most of them is an attempt to prepare for an energy-constrained future at the local level as quickly and as effectively as possible. There are many possible models to follow. Here are three different types.

Plymouth Area Renewable Energy Initiative
In 2004, in response to local concerns about the lack of planning on the national and state level for peak oil, a small group of activists in New Hampshire founded the Plymouth Area Renewable Energy Initiative (PAREI). The group, a membership-based 501(c)3, decided to focus initially on solar hot water, according to Steve Whitman, a PAREI board member. "Solar hot water was a reasonable size of project for residents; you can teach people about it fairly easily, and they can become directly involved in their energy futures," he says. Part of this initiative included Volunteer Solar Energy

Raisers, somewhat like barn raisings of the past, where community members helped install solar hot water systems on one another's homes.

PAREI's focus quickly expanded beyond solar to include a wide range of other activities. The group now also offers Energy Exchanges, potluck events held throughout the year where local residents can connect and learn from one another on a wide range of energy-related topics. In addition, an Energy Advisor Network matches members' questions with other members' knowledge in a wide range of energy and efficiency subjects. "These events have become opportunities for all of us to learn together and to share; it's built a lot of capacity in the region," says Whitman, who is also the founding chair of the Plymouth Energy Commission.

But the group soon realized that not everyone had a good site for solar or an interest in solar hot water. Nevertheless, people were coming to PAREI gatherings and wanted to be involved. "That's when we started doing the energy audits and house warmings," Whitman says. This led to training for a staff member and other volunteers to help them become certified building auditors so they could perform blower door tests, thermal camera imaging, and other energy-efficiency audit activities for PAREI's residential and commercial members. Today PAREI has more than four hundred families and businesses in fifty towns as annual dues-paying members, as well as one full-time and three part-time staff governed by a seven-member board of dedicated volunteers. The group even administers a local foods buying website known as Local Foods Plymouth.

PAREI has been so successful that other communities from the region and across the country have asked for a template to follow for their own initiatives. In response, PAREI developed a media tool kit to help others recreate the PAREI model. They now offer an introductory video ($36) that includes discussions by the founders about the thinking and actions that went into the local energy initiative; solar thermal basics; and a description of the Energy Raiser model. When other communities are ready to move ahead with their own initiatives, they are encouraged to become a Community Partner ($150), which involves a lot of additional support as well as video and documents. They are also assigned a PAREI partnership advisor who stays in touch via phone and email to provide advice and encouragement.

"This wasn't what any of us had intended, we were just trying to make our own region better equipped for the future," Whitman says. "But the fact that it's a model that other people can use to start their own initiatives is probably the most exciting part of this."

An initiative like PAREI always involves a lot of hard work and unquestionably has its challenges. Maintaining membership levels, funding and capacity issues, as well as trying to avoid volunteer and staff burnout have been some of the biggest challenges, according to Whitman. For more information or to become a member, visit the PAREI website at www.plymouthenergy.org.

Acorn Renewable Energy Co-op

In September 2005, inspired by a peak oil conference in Ohio, I returned home to Addison County, Vermont, where I discussed the possibility of forming a local peak oil and global warming response group with a number of area residents and friends. As a first step, we decided to show the peak oil film *The End of Suburbia* at a small gathering to see what would happen. A lively discussion followed, and this eventually led to more meetings and ultimately to the formation of what came to be known as the Addison County Relocalization Network (ACORN). In order to keep things as simple as possible, we initially organized as an unincorporated association. The ACORN Network, as it is now called, generally followed the community peak oil response model provided by the Willits Economic Localization (WELL) group in Willits, California. We quickly adapted a version of the Willits plan as a template for our own activities to avoid wasting a lot of time trying to figure out what to do. The main difference, however, was that we expanded the area of our activities to all of the twenty-three small towns and cities in Addison County.

One of the first things we did was form an energy committee and a food committee, the two main pillars of relocalization in our view. Other committees soon followed, but the food and energy committees remained the primary focus. The energy committee conducted a comprehensive survey of energy use in the county with the help of a Middlebury College senior. That study estimated that more than $130 million was spent on energy across the county in 2005, and that very little of this sum benefited the county's (or the state's) economy. We felt it was important to know where we were in terms of energy use if we were going to try to plan for making a shift to renewables. At the very least, the study demonstrated that switching even a small portion of those energy dollars to local energy sources had the potential to give the local economy a large boost.

The study also demonstrated just how vulnerable the county was to energy price increases and supply disruptions that were completely out of our control.

Fairly early on, the energy committee had plans to form a renewable energy cooperative that would act as a facilitator for a wide range of renewable energy and conservation initiatives throughout the county. There were already a number of successful cooperatives in the county, and we liked the one-member, one-vote model. In 2008, after a lot of preparation, the Acorn Renewable Energy Co-op was formed as a separate legal entity and was officially spun off from the ACORN Network. (The ACORN Network subsequently reorganized as a 501(c)3 nonprofit and shifted its main focus to local food.)

In October 2008, in response to soaring oil prices, the co-op began its first initiative, the importation of sixty-seven tons of wood pellets from out of state for area residents (at the time, there were no in-state manufacturers of wood pellets). As part of the co-op's goal to strengthen the local economy, we worked closely with a new local pellet manufacturer, Vermont Wood Pellet, LLC, that was setting up a new pellet mill in nearby North Clarendon, and since 2009 we have made hundreds of tons of the company's super-premium pellets available to our members at a discount.

In addition to wood pellets, the co-op has worked out a significant discount for members on solar hot water systems from a highly regarded manufacturer. The co-op has also offered an Energy Education Program Series on a wide range of topics including solar hot water, solar PV, geothermal heating and cooling, home energy-efficiency improvements, and other related topics. In addition, the co-op has arranged discounts for members from around fifty local business partners who offer a wide range of products and services, many of which are related to renewable energy or energy efficiency.

"We've worked very hard to offer our members discounts on energy-related products and services, and access to programs that will help them make the transition from our near-total dependence on fossil fuels to a broad range of renewables while at the same time providing sufficient income to make the energy co-op a sustainable business over the long term," says Richard Carpenter, the energy co-op's treasurer. "This has been a real challenge, but we are making steady progress."

But the most ambitious energy co-op initiative so far has been a solar photovoltaic, group net-metered project, Acorn Energy Solar One. The large,

FIGURE 5-1. Ribbon-cutting ceremony for Acorn Energy Solar One in Middlebury, Vermont, on January 6, 2012. Photo by Laura Asermily.

150-kilowatt project comprising 528 PV panels is a collaboration among the Town of Middlebury and two local cooperatives (the Acorn Energy Co-op and Co-operative Insurance Companies). The Town of Middlebury supplied the site (which was otherwise-unusable land at a decommissioned sewage treatment plant), Co-operative Insurance Companies provided the equity investment, and the Acorn Energy Co-op developed the project and provided some of the funding from its members. A new entity, Acorn Energy Solar One, LLC, is managing the $670,000 project and also secured the primary debt financing from the Vermont Economic Development Authority and a local bank (the National Bank of Middlebury). The three main project participants each receive a third of the electricity produced. The Acorn Energy Co-op's third of the power is further subdivided into sixteen separate subscriber accounts for individual energy co-op members. The community solar array was completed and went online in late December 2011. The total project time from early discussions in June 2010 to completion was eighteen months. Actual construction was only thirty-nine days. The energy co-op is eyeing another local site with potential for a much larger group net-metered PV project.

In its short history, the Acorn Renewable Energy Co-op has accomplished a lot. Membership has grown to around two hundred, but it has not been easy. The co-op relies on one (very) part-time staff worker and an elected, all-volunteer, nine-member board for direction and carrying most of the workload. The co-op has faced the same sort of challenges encountered by Plymouth Energy in neighboring New Hampshire: maintaining membership levels, funding and capacity issues, as well as trying to avoid volunteer and staff burnout. For more information or to become a member, visit www.acornenergycoop.com.

Snohomish County Public Utility District

The Snohomish County Public Utility District, headquartered in Everett, Washington, is the second largest publicly owned utility in the state. It serves around 320,000 electric customers and 20,000 water customers in all of Snohomish County and Camano Island. In 1946, the Snohomish County PUD began operations mainly in the water distribution business. Three years later, the PUD became primarily an electric utility with the purchase of the local electricity distribution system from Puget Sound Power & Light Company. In the early 1980s, the PUD built its own hydroelectric generating facility, and also owns a co-generation facility that burns wood for fuel.

Snohomish is one of the best examples of a progressive public utility that has put conservation and renewable energy at the top of its priority list. Hydropower accounts for about 80 percent of the PUD's electric portfolio (primarily from the Bonneville Power Administration), and several additional small hydro projects have recently been added to the PUD's holdings, with more on the drawing board. In addition, the PUD has contracts for wind energy from three facilities in Washington and Oregon and a contract with a landfill gas facility in Klickitat County. The PUD has also emerged as a leader in tidal and geothermal energy research, with $12.5 million in federal funding secured to date.

"We have no ownership or purchase arrangement for fossil fuel projects, and even though we have added a significant amount of wind, we have not—like other utilities—added natural gas," says PUD general manager Steve Klein. The PUD has also completed a fiber-optic project as part of a multi-year effort to upgrade its electric grid with smart technology. "These technology upgrades will provide the ability to measure, monitor and control power via robust two-way communications. All of this work is moving us toward the ability to support smarter systems," Klein adds.

The Snohomish County PUD is one of the first utilities in the region to adopt an official climate change policy and implementation strategies. That policy includes reliance on conservation and a diverse mix of renewable technologies including, but not limited to, wind, tidal, solar, biomass, and geothermal to meet future needs.

"We have a long history of commitment to conservation," says Klein. "In our climate change policy, we put a strong focus on conservation, renewables, and sustainability. And not only do we talk about why this is important to us, we also describe how we will walk the talk ourselves and how we have approached this with our own assets and facilities. In addition, we have put an emphasis on local power to the extent that we can develop resources in our own backyard."

As part of that policy, the PUD offers cash incentives or low-interest loans with its Solar Express program, implemented in 2009, for residential and business customers who want to install solar photovoltaic or solar hot water systems in their homes or businesses. To date, the program has resulted in more than a hundred solar installations. "The goal of the program was to break down barriers," Klein says. "The experience we had was that unless you were a retired Boeing engineer, the idea of sitting down and selecting a solar panel technology, determining what kind of inverter you needed, or perhaps a battery storage component, was overwhelming. For some people, just trying to sort out all the grants and incentives from the federal government or the state was just too much. So, what we wanted to do was simplify all of that and transform it into a turnkey project."

The PUD also offers free workshops and expert advice, a list of registered designers and installers, and assistance with Washington State's energy production incentives. The PUD also sponsors a free annual Snohomish County Solar Homes Tour featuring a dozen or more homes throughout the county and Camano Island. The self-guided tours give local residents a chance to view the projects and to discuss them with the homeowners.

In addition, customers who use solar, wind, hydro, biogas from animal waste, or combined heat and power technologies (not more than one hundred kilowatts) are normally eligible to participate in the PUD's Net Metering Program. This allows participants to connect their generating equipment to the PUD's distribution system and to sell any excess electricity that they generate back to the utility.

In another related initiative called Community Power!, the PUD worked with selected communities across the county for two years starting in early

2012 to raise awareness about conservation and to encourage community members to reduce their energy use and save money. In collaboration with the county and the City of Everett, the program provided free conservation measures and energy assessments as well as additional opportunities for reduced energy consumption through enhanced financial incentives for energy-efficient upgrades to homes and businesses. "Community Power! allowed us to build on our successful energy-efficiency efforts in homes and small businesses," says Klein. "It helped our customers save on energy costs, raises awareness of conservation in our communities, and provides additional jobs in the region."

Yet another example of the PUD's commitment to renewable energy is its Planet Power program. The program is an opportunity for PUD residential and commercial customers to support local renewable energy generation in Snohomish County above and beyond what the PUD is already doing. Planet Power supports small-scale solar demonstration projects on buildings such as schools, libraries, city halls, and other community sites. Through the program, customers can make onetime or ongoing contributions to a fund that provides grants to carefully selected sites in the PUD service area. The program is voluntary, and contributions can be as little as $3 per month (reflected on customers' monthly electric bills). When enough funds accumulate, the PUD sends out Requests for Proposal for small-scale renewable energy projects located within the county. The submitted projects that best meet the goals of the program receive grants.

The Snohomish County PUD has unquestionably initiated a lot of successful programs in recent years. "But I think our biggest success is that we have got community support for our programs and overall strategy," Klein says. "That's important, because as a public utility our ratepayers are also our owners."[20] For additional information visit www.snopud.com.

The next eight chapters continue the discussion of community energy with a series of case studies focused on various types of local energy projects relying on the many renewable energy technologies available.

FIGURE 5-2. A solar PV project at Snohomish High School that the PUD helped facilitate.
Photo by Snohomish County Public Utility District.

6

Solar

Community solar power can offer unique benefits in the expansion
of solar power, from greater participation and ownership of solar
to a greater dispersion of the economic benefits of harnessing
the sun's energy. But community solar faces significant barriers in a
market where the "old rules" favor corporate, large-scale development.
—JOHN FARRELL

What a difference a few years make. When I was writing *The Citizen-Powered Energy Handbook* in 2006, obvious examples of collaborative community solar energy projects were few and far between in the United States. And many of those existing projects were limited to installations on schools or municipal buildings. But today, thanks to the spread of group net metering and other incentives, there is no lack of examples of community solar projects, and the number—and enthusiasm for them—is multiplying rapidly.

One reason for the growing popularity of community solar is the fact that these projects are relatively easy to permit and build; they're generally not challenged by the regulatory hurdles or organized opposition often associated with industrial-scale renewable energy projects. Another reason is that you don't necessarily need a home with an unshaded, south-facing roof. Don't own a home? No problem! As long as you have an electric meter, most community solar projects provide everyone an opportunity to participate. Admittedly, community solar projects are not yet possible in states lacking group net-metering laws; and these projects are still unquestionably a challenge to organize and implement. But on balance, the benefits far outweigh the challenges and are definitely worth the effort.

In this chapter we look at a community solar gardens initiative in Colorado that is helping groups in cities and towns across the country organize their own community solar projects; a first-of-its-kind community solar project in Washington that continues to grow well beyond its organizers' original expectations; and a small community solar initiative at a cohousing project in Vermont that was assisted by a new start-up Community Energy Exchange.

Solar Gardens, Colorado

The past few years have seen a proliferation of community solar projects across the nation. The first community solar facility in Colorado was built by United Power Sol Partners in Brighton in 2008 and attracted forty-eight subscribers. The largest solar garden in the nation, a one-megawatt project called SMUD Solar Shares, was installed in 2007 by the Sacramento Municipal Utilities District (SMUD). In this project, subscribers enter into a power purchase agreement with the utility.[1]

However, many of these projects have been organized by groups who were unaware of the existence of similar projects in other communities. The lack of communication among these groups has unfortunately resulted in a lot of time being wasted "reinventing the wheel." Consequently, the most significant step in pulling all of this relatively disorganized activity together into a coherent movement has been the development of the solar garden concept.

A solar garden is a co-op, partnership, LLC, or other business entity, normally with multiple subscribers, that owns a solar photovoltaic array. Each subscriber owns an interest (or purchases a long-term lease) in the power generated by the solar array and receives a credit on his or her utility bill for a proportionate share of the electricity generated. The Ellensburg Community Solar Project (see below) would qualify as a solar garden even though it was not called that officially. In some locations, these projects are also referred to as community solar farms, solar co-ops, community-supported energy, solar shares, offsite solar, community power, and so forth. Regardless of the name or its location, it's the same basic strategy.

The phrase *community solar garden* was apparently first popularized in this country by Luke Hinkle, founder and owner of My Generation Energy, a solar installation company in Brewster, Massachusetts, with two subscriber-based facilities. The term *solar garden* was used prior to that to describe a number of utility-scale projects.

FIGURE 6-1. Pictured left to right are Joy Hughes, founder and CEO of Solar Panel Hosting; Greg Ching, chief sustainability officer; Robyn Lydick, media relations; Doug Southard, solar engineer; and Richard Deem, treasurer. Photo by Solar Gardens Institute.

One of the key players in the solar garden movement has been Joy Hughes of Saguache, Colorado, the founder of the Solar Gardens Institute, a central clearinghouse for solar garden hosts, subscribers, developers, and financiers; it offers a blog and a wide range of educational resources on its website (http://solargardens.org). In addition, Hughes is the founder and CEO of Solar Panel Hosting, an independent company that works with local installers to build and manage solar gardens. "We call it a hosting service provider," Hughes explains. "We work with the local communities to give people all the tools they need to develop and manage a solar garden. That way the solar installation company can do what it does best—build more solar arrays—and then we manage the project and bring subscribers in and handle changes in subscriptions and communicate with the utility and so on."[2]

Hughes got involved when a large, commercial-scale solar project was proposed for the San Luis Valley, where she lives. "I started thinking about how we could keep more money in the local economy," she says. "So, I looked at the model of community ownership, and that was where the idea of Solar Panel Hosting came from. When I thought about that, the whole thing just hit me in a flash, so I went and set up my company on the last day of the United Nations Climate Change Conference in Copenhagen [December 18, 2009] when it had become clear that world leaders were not going to be moving fast enough and that we were going to have to do it ourselves." Solar Panel Hosting opened a ten-kilowatt prototype solar garden in Westminster, Colorado, in March 2010, and is assisting the development of at least nineteen other solar gardens in communities across the state.

Hughes is passionate about empowering local communities to take charge of their own energy futures, and sees solar gardens as one of the best ways to achieve that goal. But she admits that this is not an easy process for the average person or community. "The whole process of community solar development is fairly obscure, and there are not a lot of people who know how to do this," she says. "However, anyone can learn to develop a solar project, and we can teach them how to do it so they can put it together for their own community."

To provide some basic guidance for this emerging sector, Hughes has set up founding principles and principles of responsible solar development: working with local contractors, listening to the community, providing an open forum for all stakeholders, and supporting local, human-scale projects that promote distributed generation. "We want the people in the community to be driving the decisions in creating their projects and owning their solar panels," Hughes says. "We bring that together with the big dollar investment required and provide the bridge to something that until now might have only been used to finance a very large project out in the desert."

One of the largest issues, according to Hughes, has been struggling with the glacial pace of bureaucracies. "The biggest challenge is dealing with what I would call green tape," she says. "Given the time it takes to bring things to the legislature, the public utilities commission, and internal utility bureaucracies, it can be a very slow and frustrating process. We got the community solar gardens bill introduced in November of 2009 in Colorado, and we didn't get the rules until 2011, so it's been a two-year process right there. Some of these citizens groups have been at it even longer than that just trying to pull their sites and plans together, and then waiting for the regulations to catch up with them."

Nevertheless, Hughes is excited by the progress that has been made in the past few years. "I would say the biggest accomplishment has been setting up solargardens.org and bringing together all of the communities that have been doing this on their own without even knowing there were others doing the same thing," she says. "I think the passage of the solar gardens bill was another huge accomplishment here in Colorado; it's been a major team effort that's sparked an emergent community movement across the state."

But this movement is not restricted to Colorado. "It's moving very quickly in the United States; I feel encouraged by the developments I see in California and other states across the country," Hughes continues. "We have been approached by people in most states now who would like to do a solar garden, and I'm feeling a bit overwhelmed given the sudden influx of potential projects and finance and media attention. We are seeing a tenfold increase in activity over the same time a year ago, and I keep finding more and more of these projects every day."

When asked what advice she would give to groups that are considering a community solar garden project, Hughes responds, "Just go out and do it. Bring the key people together in your community. Find people who can work with the county commissioners or other local officials and the utilities and form your coalition. Then go to solargardens.org and we can help you with some training to lead you through all of the steps of solar development, from concept, to attracting subscribers, to actually flipping the switch. We are gathering the best practices from all around the country and the world and bringing the solar gardeners together so they can look at what others have done to see if they can apply it to their own project. We want to share these tools with all of the other solar gardeners and advance the whole movement; it's very exciting."

Ellensburg Community Solar Project, Washington

Considered to be the first of its kind in the nation, the Ellensburg Community Solar Project in Ellensburg, Washington, came online in November 2006. Originally conceived three years earlier by the Central Washington University Energy Extension in Ellensburg and the Bonneville Environmental Foundation, the project was intended by its organizers as a model that would increase demand for solar energy. They did this by allowing community members to participate directly in the project.

"This all started at an annual solar summit where we were discussing barriers to solar and why people weren't moving forward with installations," recalls Gary Nystedt, one of the project's key organizers. "The initial concept was to get the utility to put up a rack-mounting system so people could purchase panels and then receive the kilowatt-hours that the panels produced. Some of the folks at the summit said, 'Great idea—call us when you get it done.'"[3]

Nystedt was determined to get it done. He and several others refined the concept and took it back to Ellensburg city officials, who eventually agreed to authorize the project if supporters could raise $50,000 in pledges from the city's utility customers (Ellensburg's municipal electric utility, formed in 1892, is the oldest electric utility in the state). "In a period of about three months, we were able to raise approximately $103,000 in pledges and actually expanded the initial project size from a twenty-four-kilowatt system to a thirty-six-kilowatt system," Nystedt says.

Local residential and commercial utility customers were invited to collaborate with the city to help fund the project (often described as a community solar co-op). In exchange for their financial support, the members receive the kilowatt-hours of electricity produced by the project in the form of a quarterly credit on their utility bill for twenty-plus years based on the amount of money they put in. To calculate member reimbursement, the city valued the electricity at a rate set by the Bonneville Power Administration (BPA), the main electricity wholesaler in the region. In order to keep things relatively simple, the city technically owns the project, but the contributing members can sell, assign, or donate their "shares" to any other local individual or commercial utility customer. The city also retains the rights to the so-called environmental attributes—renewable energy credits that can be traded on the open market—produced by the system.[4]

City staff launched the Ellensburg Community Solar Project with a marketing effort targeted to local residents. A minimum initial contribution by each customer/member was set at $250 (the single largest contribution received was $11,000). Once the initial contribution was made, subscribers could increase their contribution at any time and for any amount. The municipal utility allows the customer to contribute toward the project up to the point that their annual solar credit zeros out their electric bill.

FIGURE 6-2. Community members inspect the Ellensburg Community Solar Project after completion. Photo by the City of Ellensburg.

Located at the west end of a popular community park a short distance off Interstate 90 (one of the state's busiest highways), the solar project is directly in the view of thousands of travelers who pass by every day. "Our goal was to make it highly visible initially to spark people's interest and promote solar technology," Nystedt says. "Long range, we would like to see systems going on rooftops and other places to make them less visible. But initially we felt high visibility was important."

So far, the project has evolved in four distinct phases. Phase one saw the installation of 120 three-hundred-watt standard (polycrystalline) solar panels totaling thirty-six kilowatts, put in service in 2006. At that time, seventy-three members signed up to participate. In addition to these local investors, the city funded the project with a grant from the Bonneville Environmental Foundation as well as funds from the BPA's Conservation Rate Credit (CRC) program.

The second phase, completed in February 2009, added 21.6 kilowatts to the project with 72 two-hundred-ninety-watt panels. Central Washington University (CWU), located in Ellensburg, contributed the majority of the funding using state grant money, while the city again used funds from BPA. The university, which represents between 35 and 40 percent of the utility's total electric load, receives credit for the power produced by phase two.

Due to the popularity of the project, the city offered a third expansion. Put in service in November 2009, phase three included 180 thin-film seventy-five-watt solar modules to allow performance comparison with the polycrystalline panels used on phases one and two. Phase three encouraged participation by local community members (individual and business ratepayers) through local outreach initiatives.

Phase four in 2010 saw the installation of an additional forty-five kilowatts of thin-film technology. The city hopes to further expand the project to a total of 165 kilowatts. In addition to solar, the Ellensburg Community Renewable Energy Park (as it is now called) is adding nine wind turbines and a new concentrating solar tracker array as part of a Northwest Regional Demonstration Smart Grid initiative, partially funded through federal stimulus dollars to generate information on solar and wind energy production.

Because a public utility—in this case, the City of Ellensburg—is the owner, the original solar project did not qualify to receive either the Washington State Production Incentive (Senate Bill 5101) payments or the Federal Production Tax Credit. Nevertheless, the city and other participants managed to come up with a financial strategy that allowed the project to proceed. In 2009, state senator Phil Rockefeller, a strong supporter of community solar, helped pass Senate Bill 6170—enabling individual members of a community solar project to qualify for the state production incentive program based on the portion of their participation in the project. Members are now eligible to receive an annual production payment of 30 cents per kilowatt-hour generated up to a maximum of $5,000 per year per member through 2020. The fact that the State of Washington has made a significant effort to encourage the spread of smaller-scale, locally owned renewable energy projects has definitely been helpful to these projects in recent years.

The City of Ellensburg's Energy Services Department maintains a simple spreadsheet that tracks contributor, investment amount, and kilowatt-hours produced by the system. The city's billing department applies the credit to each customer's bill on a quarterly basis. This on-bill crediting has been described as "virtual net metering," but since the basic reimbursement rate is wholesale rather than retail, the term does not exactly apply. Combined with the state's community solar production incentives, however, customers will see a return on their investment in less than ten years, providing them with an additional ten years of free solar power. "The bulk of the contributors are residential homeowners, although we do have a few businesses," Nystedt says. "We've had ongoing money coming in to keep the project almost constantly out to bid."

The Ellensburg project organizers have unquestionably achieved their main goal of enabling ratepayers to participate directly in a local solar project. "This was a good example of taking a simple concept and running with it," Nystedt says. "And as people embraced and supported it, we just kept on going and watched it grow. We never imagined it would grow to the size it has."

The educational outcomes of the project have been equally successful, and underscore its true community nature. Central Washington University is closely allied with the project, and university students have participated in many ways: An engineering student designed the adjustable racking system to hold the solar modules, CWU's Civic Engagement Center provided senior marketing students to design the project logo and marketing materials, and an IT student developed a project web page.

The educational benefits of the project extend well beyond the college community. Each K–12 school in Ellensburg received a renewable energy curriculum, a science kit, and a full-day solar training session from the Bonneville Environmental Foundation. The foundation also created a series of web pages for each school that allow the students to access both live and historical data from the project.

Strong local support from the utility director, the Ellensburg Chamber of Commerce, the city manager, and city council was vital to the success of Ellensburg's project. The project received unanimous approval by the council, and three of the council members personally contributed money toward the project. "The city council and the chamber of commerce liked keeping our utility energy dollars circulating in the community; that was a really important part of this," Nystedt says. "The utility said they'd rather pay their customers than BPA for the electricity, so we've had that strong community financial commitment." Some of the other key supporters of the project included the Northwest Solar Center and the State Department of Ecology. Since 2006, the Ellensburg Community Solar Project has generated more than 734,450 kilowatt-hours of electricity.[5]

Burlington Cohousing Solar Project, Vermont

Burlington Cohousing East Village, Vermont's only urban cohousing community, was completed in October 2007. Located in the City of Burlington, Vermont, the thirty-two-home community is close to stores, places of employment, the University of Vermont, and a major health care facility, and is right next to a sixty-eight-acre nature preserve. Balancing community with privacy, the homes are independent but share extensive common spaces including a living room, a kitchen for preparing shared meals, a dining room with an outdoor terrace, a car-free courtyard, community gardens, a barn, a laundry room, and more. Community members also share a vision of a clean energy future.[6]

It should come as no surprise, then, that in 2009 a committee began to explore the possibility of installing a solar PV array to supply electricity for the common building. "We have flat roofs on most of our buildings and some of them are quite inaccessible, so they couldn't be used for a deck or something like that," says Don Schramm, the main project promoter. "We had actually put in a conduit through the roof as part of building the whole project so we could add solar later on."[7] An experienced local solar contractor analyzed electricity consumption and the available roof space on the main building and the community barn, and a group of solar arrays totaling twenty-five kilowatts was proposed. The PV system was installed on the roofs of the main building and barn in late 2010, and now meets more than 68 percent of the electricity needs of the main building.

"We have a common electric meter for our community," Schramm explains. "The lighting for the exterior and hallways and common spaces, including the kitchen, dining room, common living room, and an elevator, are all on the common meter. That is the biggest electric bill that anyone here gets, and it goes to the homeowners' association. So, it seemed to be the logical place to start, especially since we had those flat roofs available in common."

Investments in the Burlington Cohousing Solar Project were open to members of the cohousing community. There were two main goals for the project. The first was to reduce the consumption of fossil fuels, resulting in greenhouse gas emissions reductions. The second goal was to provide a profitable, stable, low-risk investment for those seeking a local, sustainable alternative to Wall Street. It was estimated that investors in the project would earn the equivalent of around a 5 percent annual return on their investment, while helping the community reduce and stabilize its energy costs and greenhouse gas emissions.

The PV arrays are net metered. Under net metering, the electricity produced during times of low electricity use by the community is fed back into the grid, causing the electric meter to spin backward and creating a full retail value credit on the community's electric bill. To avoid damage to the roof membrane, a "ballasted" rack system was used to minimize roof penetrations.[8] The 113 PV panels, warranted to last for twenty-five years, produce about 31,400 kWh per year. A system life of better than thirty years is possible with proper maintenance.

Community investors who provided most of the upfront capital ($130,000 out of a total of $154,894) became members of Burlington Cohousing Solar LLC (BCS). The minimum investment required was set at $1,000, and BCS is the legal owner of the solar system. The BCS LLC provided flexibility for the financing, allowing those in the community who wished to invest the opportunity to do so, while not burdening others who did not wish (or were not able) to invest. The LLC manages the project through an operating agreement that contains details on system management, share transfers, and distribution policies. Income, expenses, and tax liabilities are shared by all investors in proportion to their investment. Electricity produced by the system is sold to the Burlington Cohousing Homeowner's Association at 16.4 cents per kWh, increasing at 3 percent annually. The investors were able to take advantage of the 30 percent federal tax credit and receive distributions of current year earnings from their share of annual lease payments. After five years, the homeowners' association will have the option of purchasing the solar system from the community investors for its fair market value, based on projections of future energy production.

While the Burlington Cohousing Solar Project was considered a fairly safe investment, it did involve some risks due to the relatively illiquid assets and small number of investors; the potential (if unlikely) failure of the homeowners' association to make power purchase payments; and the novel financing approach. Consequently, the project was most appropriate for residents who wanted to invest their dollars within their local community, and who were pursuing a long-term goal with relatively modest returns. "People need to understand that you might get a return of maybe 4 or 5 percent over a long period of time; the payback is maybe ten to fifteen years," Schramm says. "They just need to be aware that this is not a big moneymaking venture; it's very much a 'slow money' investment. But if you want to invest in community projects, it's a great way to do it. And if you have some extra cash, it's a great place to put it."

The investment component in the Burlington Cohousing Solar Project was developed through a partnership with Community Energy Exchange L3C, also located in Burlington. The Energy Exchange helps area residents invest in community-scale renewable energy systems in Vermont, while expanding the state's distributed energy network and increasing community participation in renewable energy projects. The exchange helped to structure the lease relationship between the BCS LLC and the cohousing homeowners' association, as well assisting with many of the other technical and legal issues among the community, the installer, the local utility, and the state. In addition to a $5,000 project initiation fee, the exchange receives 5 percent of BCS net income as a fee for its ongoing account management services. The Energy Exchange used the Burlington Cohousing Solar Project as a pilot to test its financing model. "The cohousing solar project was really essential to launching the Energy Exchange," says Noah Pollock, project director. "We had been working through various models on paper for months, and we finally came to the realization that the only way to really fill out the details was to find a community group that was interested in working with us to actually try them out. Burlington Cohousing was a great partner in this effort. It's been really rewarding to see the solar project come to life and it's been great working with them."

Schramm is excited about extending the benefits of solar power from the community's common spaces to its individual homes with a group net-metering project. "What I'd really like to get started is a group net-metering project because we have more roof space available on the town houses," he says. "I think that's just going to take more time to explain to people. There is at least one family that really wants to do it, and that may be the way that it gets started. I have another member who wants to buy just one panel, and with group net metering we could do that too as long as we have enough other panels to make it work. Later on, we can add more panels. I think that will ultimately be the more exciting of the two projects."

Overall, the solar project has performed as projected, and Schramm is pleased with the results and greater energy resilience for the community. "I feel great about it," he says, "and I love the conversations that I have with people about the project. There are people who are very excited about it, who go online and look at how the panels perform, and I think that people are generally pleased that we have them. I'm very satisfied with our installer, and with Community Energy Exchange; they've both been really helpful and I'm looking forward to doing the next project."

Wind

*The greatest barrier to enacting
wind energy programs on a community scale
is not a lack of fruitful sites or effective technology,
but a lack of communities establishing a vision
and plan to use their natural resources
in such a way as to maximize
the benefit to the local economy.*
—THE MINNESOTA PROJECT

A growing number of community wind projects around the country are "community wind" in name only. While these large commercial, utility-scale wind farms do bring quite a lot of money into the (often impoverished) rural communities that host them, the vast majority of the money generated by these projects is funneled out of town—and often out of state as well. With most of these big wind farms, the ownership of the turbines also remains with the corporate owner/developers, offering little additional benefit or incentive for the host community that has to live with dozens (or hundreds) of huge turbines in their backyards. This does not mean that these large projects are inherently bad; rather, it means that these communities could do a lot better for themselves and their immediate environment if they instead pursued the local alternative.

This chapter profiles a small wind developer that is pioneering a true community wind initiative in Washington based on a virtual net-metering strategy; a municipally owned community wind project in Minnesota that was able to successfully use a less-than-perfect site; and a cooperatively owned commercial-scale wind farm for two small island communities off the coast of Maine.

Cascade Community Wind Company, Washington

As a professional in the wind industry for more than ten years, Terry Meyer has given a lot of thought to community wind issues. Meyer, who lives in Washington State, has worked with developers, banks, and utilities to build a number of large-scale wind projects. He has firsthand knowledge of how these projects work, and where the financial benefits actually flow.

Meyer evaluated the potential benefits of local community-owned wind projects and found that the rules, regulations, and incentives all favored large commercial projects. Then in 2008, the local utility changed its payment offering for distributed renewable electricity generation (less than two megawatts). That, combined with changes in state and federal incentives for smaller-scale local energy projects, led Meyer and several others to decide that the time was right to form Cascade Community Wind Company (CCWC), with headquarters in Bellingham. "The incentives used to be something that only large corporations could take advantage of, but now regular people could use them as well," Meyer says.[1] At the time, Washington had no community wind projects.

The main idea was to find appropriately windy sites, test the wind resource, and then engage the local community in raising funds to put up the turbines, benefiting everyone involved in the process. Meyer and his partners began to develop a new model for small community wind projects that involved some innovative investment, ownership, and power subscription models. "We work with landowners to lease their property using a fairly typical wind lease, but we share the ownership in the turbine we put up with them," he explains. "Investors in the project get a preferred return but have a low overall ownership percentage, while Cascade Community Wind and the landowner have a lower financial benefit but a higher ownership level. So we keep control of the turbine while also giving our investors a decent rate of return." Virtually all of the investors are "local mom-and-pop types," according to Meyer. In a typical project, CCWC will own about 60 to 70 percent of the turbine, the landowner perhaps 10 percent, the investors another 10 or 20 percent, and the balance might be owned by a professional who has put some sweat equity into the project. It's a collaborative venture where everyone wins.

Then there are the subscribers. "The subscribers are really the heart of all this," Meyer says. "A good part of the subscription model is our own innovation, but we also encountered the Ellensburg Solar Project that uses a similar model and we made some refinements to ours based on how they did it. Because we get paid a decent rate for the power, we can use a form of virtual net metering, based on payments rather than kilowatt-hours credited to a subscriber's power bill. In this model, the subscribers put up money for one of our turbines, but they also save money on their power bill over the next twenty years. Every time we get a check from the utility for the power we produce, we apply a credit to the subscriber's electric bill according to their percentage subscription in the project. This also allows us to have subscribers from different utilities, and there are four different utilities that we are working with." Because these subscriptions are virtual net metered, the amount of the credit from a subscription cannot exceed the amount of the subscriber's total annual electric bill.

CCWC is also following an innovative strategy for its turbines, using remanufactured 120-kilowatt Vestas V20 turbines that had been replaced with newer, larger turbines in recent years by the wind farms that previously owned them. CCWC had originally planned on using even larger turbines, but quickly discovered that getting local permitting for them was going to be a challenge due to their height, and opted to go for the smaller V20s instead. "We put our plans for larger turbines on hold and started a second project with another V20 so we would have two projects in parallel," says Meyer. "That way, if we had a delay with one, we had a second project to work on. In the end, both of them made it to the finish line." But getting to the finish line proved to be quite a challenge.

"We ran into a world of issues on these first projects; there definitely were some growing pains," Meyer continues. CCWC originally planned to have the first of its turbines up and running in less than a year. The process actually took nearly three years due to a wide range of problems. The utility originally insisted on a grid protection system that would have cost more than the turbine. In addition, there were no experienced turbine installers in the area, so CCWC decided to train a cell tower installer, leading to cost overruns. The turbine remanufacturing company was slow in filling the initial order, further delaying the project. "Pretty soon it wasn't winter anymore with hard ground where we had planned to set the turbine down, but rather it turned into a springtime, wet, muddy mess. So, we ended up spending $10,000 on gravel and bark just to be able to get into the site, and that delay resulted in more cost overruns," Meyer says. "Our V20s ended up costing us about $400,000 per turbine, which was $100,000 more than we had expected. Putting up a 120-kilowatt turbine should not be a three-year project; we were expecting it to be a six-month project, and it will be, once we have the kinks worked out."

In February 2010, CCWC signed power purchase agreements with Puget Sound Energy for the output from its first eight community-supported wind turbines. In July 2010, CCWC received a $47,975 grant from the U.S. Department of Agriculture Rural Energy for America Program (REAP) to help with the construction of the first turbine. The following month, CCWC was awarded a $1 million grant and loan (30 percent grant, 70 percent low-interest loan) from the Washington State Energy Program to assist in its community wind turbine initiative. According to CCWC, the funds will leverage approximately $10 million in private and federal funds, assuming they are allowed to install their bigger turbines. Of particular interest to the Washington State Energy Program were CCWC's efforts to remove barriers to distributed community renewable energy projects. These projects are still relatively new to Washington State, and often encounter many barriers: unhelpful permitting codes and utility policies; and lack of public education, a trained local workforce, or access to insurance and bonding.

The first CCWC project to come online was appropriately named "First Up!" The turbine, with a sixty-foot-diameter rotor atop an eighty-foot tubular tower, is located on a ten-acre private site near the town of Thorp and began production in June 2011. The turbine should produce enough electricity to power about forty-five energy-efficient homes, and CCWC estimates that each $3,800, 1 percent subscription will yield approximately $4,700 in electricity bill savings over twenty years. Two months later, CCWC completed installation of their second community-supported turbine at the 3 Bar G Ranch, a cattle operation on Thorp Prairie. A 1 percent pre-construction subscription on this turbine cost $4,400 and is projected to create $6,000 in power bill savings over its twenty-year life. In September 2011, CCWC announced that it had received $128,000 from the REAP program for a third turbine. This turbine will also be located on the 3 Bar G ranch near the existing turbine on that site. The two turbines will share a power line and an access road, which should help speed the installation process for the third turbine.

From the first turbine location in Thorp, there is a view across the Yakima River of the hundred-megawatt Kittitas Valley Wind Farm (KVWF) with its forty-eight 2.1-megawatt turbines atop four-hundred-foot towers. The comparison between the two projects is stark—and instructive. "That big wind farm pays taxes, and it also pays a lease to the landowners, and the national operation and maintenance company employees live nearby, so that money at least goes into the local economy," says Meyer. "But for every dollar KVWF gets for their energy, the absolute maximum that ends up staying in the community is maybe 10 percent."

"But with CCWC's little turbine, the financing is local, the ownership is local, the subscribers are local, the maintenance is local—virtually every bit of this picture is local," Meyer continues. "More than 90 percent of our revenue stays local, so we have a ten-to-one multiplier in the economic benefit that accrues to those living in the area of our community wind turbine. The bottom line is that the money stays here along with the turbine. That big wind farm is putting a lot more money into the economy because it's so much bigger, but if I were able to put up ten megawatts of community-owned wind turbines as opposed to the hundred megawatts of the big commercial project, all of a sudden I'm having the same local economic impact as the much larger commercial project. So, if we can get the same amount of economic benefit locally from a project that is a tenth the size, it seems obvious to me that we should be pursuing community wind just as vigorously, if not more."

Nevertheless, Meyer is quick to emphasize that he does not see the large commercial wind farms as competitors: "They're just another piece of the big picture. We need all of the renewable energy we can produce." The large commercial project across the river endured years of strong opposition from area residents. By comparison, CCWC's smaller projects have been welcomed by most of the neighbors. "I've got this glowing letter from a neighbor who lives right next to one of our turbines, and he's just thrilled about it," Meyer says. "We've given him half a subscription along with the other immediate neighbors; we feel that he should be benefiting from it too. He's getting a little bit off his power bill, but he is just very excited to see it happening. Another neighbor wanted to know if he could have one of our turbines on his property too. That's the kind of response we've been getting from people who have had direct experience with our projects."

Although Meyer wishes it hadn't taken so long to get CCWC's first turbines up and running, he's pleased with the results so far. "I think making that first leap from concept to functioning project is awesome," he says. "The whole point is that a local energy economy is beneficial to everyone involved. I've worked on mega wind farms for some big companies, but to actually put the power in people's hands to let them create their own local wind power solutions is something I'm very proud of."

City of Willmar, Minnesota

The City of Willmar (population 19,610) is located almost a hundred miles west of Minneapolis. Willmar's community wind program, which is co-op based, actually began as a proposal for a private business venture. In December 2005, Jon Folkedahl, president of an energy consulting firm based in Willmar, met with representatives from Willmar Municipal Utilities (WMU). Folkedahl proposed building his own turbines in Willmar and selling the energy they generated to the utility. WMU, which serves more than ninety-two hundred electric customers, decided that an investor-owned project like the one he proposed was too costly and would drive up energy costs for its customers. However, WMU found the idea of investing in wind energy appealing, and asked Folkedahl if he would be willing to build the turbines for the municipality instead. This would essentially cut out the middleman (in this case, Folkedahl himself).[2]

One of the benefits of this model of community wind is its cost efficiency. Capital invested in the project to cover construction and generation costs only needs to be matched by money generated through energy production. After building and operating costs are recovered, all additional money earned translates into savings for consumers, rather than profits to satisfy shareholders. This is an excellent funding model for community wind, as the cost to maintain turbines is relatively low, and energy generation is virtually free.

Folkedahl agreed to oversee the project, and in May 2006 Willmar officially hired his company as its wind energy consultant. Folkedahl conducted a feasibility study and analyzed the various risks, costs, and permitting concerns involved with the project. The Willmar Municipal Utilities Commission reviewed his study and voted to proceed.

In September 2006, a crucial phase of the planning began. Because potential wind capacity in Willmar is not particularly high compared with other places in the state, the project depended on two things to make it viable. First, siting for the turbines needed to be perfect, according to Folkedahl, in order to make up for a general lack of high-velocity wind in the area. Furthermore, because the turbines had to be erected within WMU's service territory, potential locations for the project were confined to the city limits; finding a site with adequate wind capacity within such a relatively small area was even more challenging. Second, implementation needed to be completed in as cost-effective a manner as possible. What the project would lose in wind capacity, it had to make up for in efficiency. In most projects, the wind developer directs the entire process, acquiring the turbines and overseeing construction. But to achieve the low price point needed by the utility, Folkedahl decided to find the turbines himself and hire subcontractors to complete the construction, while he represented Willmar as the project's general contractor.

The city and Folkedahl spent the next year negotiating a land lease deal with the Minnesota State Colleges and Universities System over a potential site for the turbines located within the boundaries of Ridgewater College in Willmar. Just as the two parties appeared close to finalizing the agreement, the Minnesota Department of Transportation stopped the project, due to concerns over how the turbines might interfere with air traffic patterns to and from Willmar Municipal Airport. Disappointed but undeterred, Folkedahl turned his attention to another location on the opposite side of the city, adjacent to Willmar High School. The new site proved satisfactory to both the city and the school district, which owned the land. The location had good wind exposure, although not as good as the first site, according to Folkedahl. But it had the added benefit of being within sight of the school, which gave teachers the opportunity to integrate the turbines and wind energy generation into their curricula (which they ultimately did).

In February 2008, Folkedahl called for bids from manufacturers on a pair of two-megawatt turbines. There were no responses. The United States was still in the midst of its wind farm building boom, and demand for turbines was high. Also, Willmar required a bid bond from any manufacturer interested in selling its turbines to the WMU. According to Folkedahl, the idea of entering into a public bidding process when the supply-and-demand equation was tipped so drastically in favor of manufacturers simply didn't make sense to anyone selling turbines at the time. "It was absolutely a seller's market, and the manufacturers felt no need to extend themselves for a two-turbine sale," Folkedahl recalls. "There was also a lack of certainty that they would be a successful bidder, and they just didn't want to invest any time and energy into it."[3] Willmar was forced to put its wind program on hold.

Folkedahl continued his unsuccessful pursuit of the turbines until May 2008 when Dave Laursen, president of Windations, a wind development company based in Willmar, approached him. Laursen had heard of WMU's trouble securing turbines, and he wanted to help. By becoming a turbine vendor, he could expand Windations's share in the wind business while helping Willmar acquire the turbines it needed for its project. Laursen entered into an exclusive supply agreement with DeWind, the turbine manufacturer. He agreed to serve as an intermediary between WMU and DeWind, thereby guaranteeing delivery of the turbines and eliminating Willmar's need for a bid bond from the manufacturer.

Folkedahl reopened the public bidding process in October 2008, this time calling for offers from excavation companies experienced in large-scale concrete foundation projects. This round received a considerably larger response than his initial request for manufacturer bids. Half a dozen construction companies submitted bids for the contract. The following month, Folkedahl and the WMU selected a local excavating company with the capacity and machinery to complete such a large project.

Construction on the 256-foot, twin DeWind two-megawatt turbines began almost immediately, and was completed in August 2009. There was a dedication ceremony on September 3 celebrating the completion of the first two wind turbines in the state to be owned by a municipality. WMU officials expected Willmar's turbines to provide up to 5 percent of the city's energy. Folkedahl's projections were a bit more conservative—closer to 3.5 percent. As it turned out, the figure was almost ten million kWh, or about 3 percent, due in part to unexpected problems with the turbines' sophisticated electronic controls that led to downtime for repeated repairs (fortunately covered by the manufacturer's warranty). Nevertheless, community wind will make a significant difference in Willmar. Not only will the city produce around 236,000 tons less carbon over the life span of the turbines, but Willmar

FIGURE 7-1. City of Willmar, Minnesota, community wind turbines. Photo by Willmar Municipal Utilities.

residents can expect to see a decrease in energy costs over the next twenty years. And the city is one step closer to energy self-reliance.

Willmar's community wind project had a lot going for it from the beginning. It depended on a single local funding source, it was managed by a local project supervisor working together with the municipality, and, because Willmar is served by its own municipal utility, there was no need to sign a power purchase agreement with an outside energy provider. Today all the electricity generated by the community wind project goes to power the homes, businesses, and municipal buildings of Willmar. The project experienced relatively little political difficulty and a minimum of waste. Because of the absence of a for-profit corporation, there was no stipulation that the project make money for investors, only that it meet the community goals of creating meaningful savings for utility customers and reducing the community's dependence on imported sources of energy.

This is very different from a typical business model, Folkedahl notes. "One of the more important aspects of community wind is to recognize that it should not be viewed as a business venture," he says. "Throughout this process I have been reminded that the business community has a far different outlook than those who would support community wind. The business community in general, and I'm not finding fault with them, has to look for short-range returns because the future is far too uncertain. Community wind is not a business investment, it's a long-term investment in the community's future. It's just a different game. These wind turbines, regardless of the current slowdown in the economy and energy prices, will still be an excellent community asset in the long run."

Beyond generating clean electricity, there is another way in which the wind project has benefited the community, according to Folkedahl. "I think perhaps the biggest success has been the visual impact of the project and the impression that it generates about the community," he says. "The turbines are in a very visible location; when you come to Willmar, you can't help but see them. I think that they convey the message that Willmar is a progressive community, that it's investing for the future, and that it cares about the environment."

Folkedahl believes there have been a number of key lessons learned from the project. "We learned that community involvement from the beginning was crucial. We identified the need for public support early on and we met that need by holding numerous meetings throughout the community. I didn't hear one single negative comment generated during that entire three-year time span when we were developing the project," he says. "We also learned that the romantic concept of renewable energy is not necessarily synonymous with the reality of actually producing it, but we learned that the effort was worthwhile."

Fox Islands Wind Project, Maine

The Fox Islands of North Haven and Vinalhaven, located in Penobscot Bay about twelve miles from Rockland, Maine, are two of only fifteen year-round island communities that remain from the hundreds of communities that were active off the Maine coast in the nineteenth century. Today the electric rates on these islands are anywhere from double to seven times the national average. It should come as no surprise, then, that the Fox Islands Electric Cooperative had an interest in developing the island's considerable wind resource and began to actively explore the possibility in 2001. The following year, the cooperative received a grant for a three-year wind-resource study conducted by the University of Massachusetts Renewable Energy Research Laboratory. The land used for the wind-measuring anemometer was located near an abandoned quarry in the middle of the island and on property owned by two residents who were willing to lease it on very reasonable terms for a possible wind turbine site.

While the cooperative was gathering data on the islands' wind potential, its undersea electrical cable to the mainland was becoming increasingly unreliable. Through a grant and loan from the USDA Rural Utilities Service (RUS), the cooperative received funding for a new cable, installed in May 2005, that provided a reliable way for power to be imported to the islands or exported back to the mainland. The new cable turned out to be vital for the development of the community wind resource because when the proposed turbines generated more electricity than could be used on the islands, it could be sold to the mainland, and when the wind was not blowing, power could be purchased from the mainland.

In early 2008, the cooperative requested assistance for the wind power project from the Island Institute, a nonprofit located in Rockland, Maine, that helps coastal communities remain vibrant places to live and work. The institute worked with George Baker, a member of the Harvard Business School faculty, to complete a preliminary economic analysis for the Fox Islands Cooperative. This analysis included a financial model suggesting how the island communities might finance their wind power project by finding a tax-equity investor and selling excess power via the co-op's new underwater cable.

A separate for-profit legal entity, Fox Islands Wind LLC (FIW), was formed to take advantage of the tax credits, with Baker selected as CEO. He and the Island Institute secured the commitment of a Maine company for an approximately $5 million investment in exchange for the tax credits. Permanent financing was obtained via a loan from the USDA's RUS program—the first time a RUS loan was made to a wind project on a project finance basis.

The Island Institute and the electric co-op then funded a logistical feasibility analysis, an environmental study, and a visual impact simulation. The study concluded that the wind turbines could be delivered by barge and erected for a total of approximately $12 to $14 million, with minimal impact on birds, wetlands, rare plants, and habitats. A detailed economic analysis suggested that the project could lower electric power rates by 2 to 4 cents per kilowatt-hour for the first ten years and twice that amount during the second decade—representing a substantial savings for co-op ratepayers.

After an extensive public education and outreach campaign coordinated by the Island Institute, in July 2008 the ratepayers on both Vinalhaven and North Haven voted overwhelmingly (382 to 5) to authorize the Fox Islands Electric Cooperative Board to proceed with developing detailed plans to erect up to three turbines on the site used for the earlier wind measurement tests. Baker and Island Institute staff then raised $350,000 in pre-development loan funds from a variety of private individuals and foundations to finance additional engineering, legal, and environmental information required for project permits. FIW also partnered with the Island Institute to raise pre-development capital for attorney fees; for interconnection and environmental studies; to complete necessary local, state, and federal permitting; and to secure permanent financing for insurance, operations, and maintenance.

Site construction began on June 8, 2009. Foundation work took place throughout July, and turbine components began to arrive on the island in August, after having been barged across Penobscot Bay. "I understand that there were more than a hundred people down on the wharf watching the first turbine components come off the barge and they all erupted into applause when the truck left the wharf to haul them up to the site," says Suzanne Pude, community energy director at the Island Institute.[4] The first tower section was erected on September 1, and by the end of that month all three turbines had been installed. The project continued on schedule as electrical work and grid tie-in took place in October. The turbine manufacturer, General Electric, arrived on the site in early November to begin the commissioning process for the turbines. On November 17, 2009, almost five hundred islanders, politicians, and members of the press gathered at the project site to celebrate the successful, on-time completion of the 4.5-megawatt project; on December 1, the three turbines officially began commercial operation.[5]

FIGURE 7-2. Fox Islands Wind Project, Vinalhaven, Maine. Photo © 2011 by Peter Ralston/Island Institute.

"It is exciting to see the turbines in operation," says Baker, who attended the celebration. "To see them providing benefits to these islands is the culmination of years of contribution by many people. These island communities are truly leaders in the field of coastal renewable energy."[6] Overall, the project cost $14.5 million, just slightly above the original estimates.

However, there was a small group of local residents living close to the turbines who raised concerns about the noise created by the spinning blades under certain wind conditions. There are thirty-eight households located within half a mile of the turbines, and approximately half of them are seasonal homes. At least half a dozen neighbors say that the noise is extremely disruptive, while other year-round residents who live within the half-mile zone have not complained or do not find the noise to be a problem, according to Chip Farrington, the Fox Islands Electric Cooperative's general manager. In response to the complaints, the cooperative ran a series of tests that randomly slowed the turbines at night and asked the neighbors to keep detailed logs of the sound levels during the tests. While those tests were inconclusive, they did provide useful information to the cooperative and others who were trying to find ways to mitigate the problem.

The situation became more complicated when the unhappy neighbors formed a group called Fox Islands Wind Neighbors, hired an attorney, and filed more than a dozen complaints with the Maine Department of Environmental Protection (MDEP) as well as a number of lawsuits. In response, the MDEP mandated operational curtailments under certain wind conditions on the turbines during a sound monitoring process that was supposed to last up to a year. However, the monitoring actually lasted more than seventeen months, causing over $365,000 in unexpected expenses for the project to pay for lawyers and sound consultants. As a result, in April 2011, FIW notified the cooperative that a 1-cent-per-kWh rate increase (about $5.59 per month to the average residential bill) would be necessary to help cover the extra expenses.[7]

As part of the ongoing efforts to resolve the problem, a team of German specialists hired by General Electric arrived on the island in July 2011 and spent several months installing serrated edges on the first forty feet of the trailing edges of each turbine blade in an effort to test whether or not this would reduce the noise problem. This was only the third time this technology has been tried worldwide. "The comments have been very positive from people who have gone up to listen to the turbines," Farrington says. "There seems to be a general consensus that the sound has been noticeably lowered; I think we are heading in the right direction."[8]

Although they are generally sympathetic to the affected neighbors, the vast majority of the island's residents still strongly favor the wind project. In fact, that support has only gotten stronger, especially among most of the island's twelve hundred year-round residents, who are seeing an average annual savings of $300 in their electric bills. "The community support for the wind project is very strong," Farrington says. "The high electric rates we had in the past were the main driving force, and this project means long-term stability in the rates that will protect us from future surges in the energy markets."

The second goal was to give the community control of the project, and that has also been accomplished.

In its first year of operation, the wind project exceeded its original production target by over 4 percent, with 12.1 MWh of power produced, more than enough to meet the islands' needs; so from a business standpoint the project has been a great success. In late June 2011, the wind project received a favorable decision from the Maine Department of Environmental Protection, finalizing the operating protocols for the turbines that addressed the noise issue with minimal financial or operational impact on the project. In September, the Maine Public Utilities Commission rejected a lawsuit filed by the neighbors against the cooperative, finding the complaints were without merit. Clearly, FIW and the cooperative are making a good-faith effort to resolve the sound issue, and everyone involved hopes to reach an acceptable compromise, not only for the affected neighbors, but for the entire community as well.

"Admittedly, there have been challenges, but overall I get the sense that there is a lot of pride on the islands that, as a small and remote community, they were able to pull something like this off," Suzanne Pude says. "There is also pride in the fact that they are generating their own power, which is very much in line with the idea of self-reliance out in these remote areas."

Hydroelectricity

*We were looking for new renewable energy resources that had
a low environmental impact, and that's when this idea came up.
Capturing excess energy that we would normally be wasting
was a concept that we really liked.*

—Jeff White

Hydropower is not a new idea. Most community hydroelectric facilities have been around for a very long time. In 1889, there were already about two hundred hydroelectric plants in the United States, and many of them were local municipal projects. At its peak in the 1940s, hydropower provided around 35 percent of this country's electricity. Then the hydropower sector began its long decline relative to fossil fuels. Many of the smaller community facilities that had not already been abandoned during the Great Depression were closed in favor of cheaper electricity generated elsewhere.

Many of those communities have since regretted that decision as the cost of electricity has increased. More recently, however, new tax credits, grants, and other incentives to reduce greenhouse gas emissions and encourage renewable power have prompted many communities to take another look at restoring or upgrading local hydropower projects.

As mentioned in chapter 5, one of the biggest impediments to reviving community hydropower has been an enormously expensive federal/state regulatory and approval process that moves at a glacial pace. Five years is not an unusual time frame for obtaining a Federal Energy Regulatory Commission (FERC) hydropower project license. Some state approvals can take almost as long.

FERC's recently established Small/Low-Impact Hydropower Program has provided a faster track for many local hydro projects. Then in August 2010, the agency went a step farther when it signed a Memorandum of Understanding (MOU) with the State of Colorado to develop a pilot program for simplifying and streamlining authorizations of "conduit exemptions" and small (under five megawatts) projects while ensuring environmental safeguards. FERC also agreed to waive certain consultation requirements when all of the relevant resources agencies agree to do the same. FERC has subsequently signed similar agreements with a number of other states.[1]

In-conduit (water pipe) projects are some of the smallest, easiest, least expensive types of community hydropower initiatives. These projects often involve bypassing pressure-reducing valves in municipal water systems with small hydroelectric turbine/generators. Sometimes discharge pipes at water treatment plants are fitted with micro-turbines as well. There are literally thousands of potential in-conduit projects around the country that many municipalities and private companies are exploring. As we shall see, the new conduit exemptions have already substantially reduced the FERC approval time for a few of them.

While most new hydroelectric projects in this country are constructed by large investor-owned utilities or private companies, there are some exceptions. This chapter describes three different community-owned projects of a range of sizes: a micro-turbine installation in Utah that sailed through the FERC licensing process in just a few months; a small municipal project that took advantage of an opportunity to buy an existing privately owned hydro plant in South Dakota; and a very large hydroelectric project being constructed at an existing federal dam on the Ohio River for a consortium of municipalities.

Logan, Utah

Located at the base of the Rocky Mountains in northern Utah, Logan (population fifty thousand) is the principal city in the region. The city and the river that flows past it were named after Ephraim Logan, an early fur trapper. Logan is also the home of Utah State University, with around twenty-five thousand students enrolled. The city is served by Logan City Light & Power (LCL&P), a locally owned public utility incorporated in 1904. The utility owns an electricity generating station located in the city that houses three natural-gas-fired turbine generators each rated at five megawatts, as well as several run-of-river hydroelectric facilities along the Logan River in Logan Canyon. With about eighteen thousand customers, LCL&P has a peak load of around a hundred megawatts, and gets about 24 percent of its electricity from renewable sources: some from the Colorado River, some from wind turbines in Wyoming, and an increasing amount from hydroelectric facilities in Logan Canyon.

The Logan Public Works Department's Water/Sewer Division is responsible for the city's water supply. Logan draws 70 percent of its potable water supply from Dewitt Spring, located about seven miles away in Logan Canyon. From the Dewitt collection basin, a pipeline runs down the canyon 5.2 miles to a flow control vault (a concrete building containing a pressure-reducing valve), where excess pressure is removed before the water is distributed into several concrete storage reservoirs.

The original steel pipeline (or penstock) was constructed in 1934, with some sections upgraded to reinforced concrete in 1949. Over the years, some sections began to leak, so in 2008 major portions of the old twenty-four- and thirty-inch pipes were replaced with thirty-six-inch steel pipe. The larger-diameter pipe increased available water flow to the growing community while significantly reducing pipeline friction. However, less friction resulted in additional head pressure[2] at the flow control vault, creating both a problem and an opportunity. The existing pressure-reducing valve at the flow control vault was pushed beyond its design limits under the higher pressures, causing severe stress and rapid deterioration of the valve.

During the pipeline upgrade planning in 2007 an "energy recovery" approach was suggested. This involved replacing the pressure-reducing valve with a micro-turbine, thereby reducing the excess head pressure and generating electricity for the municipally owned power system at the same time. One of the biggest challenges would be to fit the new hydroelectric system into the limited floor space of the flow control vault. "This is not a new idea, but it was new to us," says Jeff White, former director of Logan City Light & Power. "We were looking for new renewable energy resources that had a low environmental impact, and that's when this idea came up. Capturing excess energy that we would normally be wasting was a concept we really liked. We also wanted to find a resource that was reliable, and this was more reliable than a wind turbine or solar array because the water runs all the time."[3]

Since the project involved the installation of a hydropower unit onto an existing water line, there were no environmental impacts to deal with, greatly simplifying the approval process. FERC had recently streamlined its permitting processes for small hydro projects with low environmental impacts, and approval came quickly in just a few months (instead of years, which until then had been the norm). The State of Utah also required a "non-consumptive" water right permit for the city to produce the power. Public support was very strong because the project produced clean, renewable power simply by recovering energy that had previously been going to waste.

The overall cost of the project was approximately $1.5 million. "We'd hoped to do it all on our own, but we just couldn't afford it," White says. "When the federal Energy Efficiency and Conservation Block Grant [EECBG] Program came out in 2009, we applied for a grant to cover part of the total project cost. Our application was approved, so we were able to do it with a fifty–fifty match." The city's part was approximately $750,000, paid for by tapping a capital reserve fund that was saved for projects like this so Logan could avoid the need for bonding.

Despite the relative simplicity of this type of project, there were a number of technical challenges to overcome due to some unique local factors. Even though large sections of the penstock had been replaced, a two-mile length of smaller concrete pipe still remained, effectively choking flow and creating variations in head and flow. Consequently, the turbine design had to operate smoothly and efficiently across the present wide range of available head, and also across a different range in the future when remaining sections of twenty-four-inch concrete pipe are replaced with thirty-inch pipe. This upgrade will nearly double the available head pressure and turbine power output.

The remaining old section of concrete pipe created the potential for an additional problem: catastrophic "water hammer"[4] events, even with minor pressure surges. Consequently, special attention had to be given to the design of valve closure operations, turbine overspeeds, and any other factors that might contribute to water hammer events. The project designers had to keep in mind that power generation was a secondary priority, since the main function of the water system was to provide water to the community. If an overpressure event caused by the new turbine destroyed a section of the water line, the entire community would suffer with rationed water until repairs could be made. Consequently, city officials had to ensure that the needs of the water department and power department did not conflict.

The project designers decided that the final line of defense for protecting the integrity of the penstock would be provided by the installation of a "surge pipe." The surge pipe would tap into the penstock and be routed up a steep hillside next to the flow control vault. During an overpressure event, a U-shaped section of piping at the top of the surge pipe would redirect the surge water over the U and back down the hillside into one of the storage tanks. Under normal operation, the water elevation in the surge pipe would remain just below the U.

In addition, even though the turbine would replace the need for a pressure-reducing valve, if the turbine shut down for any reason the city would still have to ensure water delivery by using a bypass system. With this system, when the turbine inlet valve closed, a neighboring valve at the pressure-reducing valve would have to open automatically to bypass the turbine and maintain water supply to the storage reservoirs.

The site characteristics dictated a Francis-type turbine,[5] a design normally well suited for varying flows but less tolerant of wide swings in head pressure. The challenge was to come up with a single turbine designed to handle both current and future head conditions. "When we replaced the pipeline, we only replaced the worst part of it, so what we needed to do was design this turbine for the future," White says. "At present, the restriction caused by the old pipe means that during high flow periods in the summer we don't have enough pressure, so there would be times in the summer when we would have to take the turbine offline. But in a few years, when we replace the rest of the old pipe, we will be able to run the turbine full-time, plus get more electricity than we would have if we designed it to run full-time now."

The city turned to Canyon Hydro in Deming, Washington, for the design and manufacture of a custom Francis turbine able to accommodate the vastly different head pressures. Canyon Hydro eventually came up with a design that met the utility's present and future needs, although in terms of size the turbine turned out to be a tight fit. The new flow control vault that had been built as part of the city's pipeline replacement project a year earlier was deliberately designed larger than needed for the pressure-reducing valve, with the hope of someday adding a turbine. However, even with this forward-looking planning, space remained at a premium. Despite the best efforts of the designers, the horizontally mounted 220-kilowatt turbine/generator with all its components would not fit within the building, and eventually an alcove extension had to be added to one wall—with some help from concrete saws and jackhammers.

FIGURE 8-1. The interior of the flow control vault showing the new Francis-type turbine/ generator and pressure-reducing valve. Photo by the City of Logan.

In the end, everything fit in the newly expanded space with ample room for the turbine/generator as well as control cabinets and switch gear. The grid connection was made using the same transformer serving the vault, minimizing interconnection costs. The project came online in March 2011. The system, which provides enough electricity to power around 185 homes, was installed on time and within budget except for the alcove addition onto the vault.

So far, the turbine's performance has been excellent, according to White. "It's been doing great, we're very pleased with it," he says. "I think it's the way to go, especially when you are taking a liability—the excess pressure—and turning it into an asset—renewably produced electricity—for the community. It's a win–win for everybody." Since start-up, community interest and support have remained strong.

Encouraged by the success of the micro-turbine project, the city is pursuing other potential hydro projects. There are several other locations in the water system where a micro-turbine might work, and a much larger hydro project—the Porcupine Dam at the south end of Cache Valley—is also being investigated. "It's an old hydro site that was abandoned about twenty-five years ago," White explains. "We are trying to restore it so we can manage it locally ourselves." LCL&P is even developing a local solar PV garden project available to any electric customer in the city. "We're looking at things we can do here, rather than having to rely on new transmission resources to bring in power from other distant locations," he says. The best part of all of this new emphasis on local power is that the owners, the citizens of Logan, are one step closer to securing greater energy sustainability for many years to come.

Spearfish, South Dakota

Hydro Plant No. 1, located in Spearfish, South Dakota, was built in 1910 by the Homestake Mining Company to replace a small city-owned hydroelectric plant that had been destroyed by a flood. The new plant provided electricity for the company's famous gold mine in the town of Lead and to power the Homestake Sawmill in Spearfish, which supplied timber for Homestake's mining operations. A key part of the new hydroelectric project was a five-mile, stone-and-concrete-lined aqueduct tunnel that diverted water around a geological "loss zone" in Spearfish Canyon to help maintain water flow even during drought conditions.[6] Before its closure in 2002, the Homestake Gold Mine in Lead reached more than eight thousand feet below the surface and was the oldest, largest, and deepest mine in the Western Hemisphere. In 2004, Homestake approached the City of Spearfish to see if the community would be interested in buying the hydro plant and, in particular, the diversion tunnel and water rights that came with it. The city was interested.

"While it remained in private hands during all those years, the hydro plant was never required to have a FERC license," says Cheryl Johnson, Spearfish's public works administrator. "As Homestake announced their plans to close up the mine and began to divest themselves of their properties, it was determined that they would have to go through the FERC licensing process for the hydro plant." Homestake contested that decision but eventually lost the argument. "At that point they decided that instead of trying to go through the whole licensing process themselves, they would give the city a chance to buy the plant."[7]

FIGURE 8-2. Interior view of Hydro Plant No. 1 in 1911. Photo by the City of Spearfish.

After several months of negotiations, a deal was struck between Homestake and the city for $250,000. The purchase included the hydro plant, a small intake dam in Spearfish Canyon, a forebay pond, two 1,200-foot-long wood-stave pipelines, two mile-long steel penstocks, four 54-foot-high surge towers, and (most important of all to the city) the five miles of water diversion tunnel and Homestake's water rights to Spearfish Creek. The hydro plant contained a pair of two-megawatt generators that could power about two thousand homes.

"The hydro plant is adjacent to our city-owned campground park area, and the plant is tied to the tunnel that allows Spearfish Creek to bypass the loss zone area," Johnson says. "So our interest really was in maintaining the creek flow the way it had been for the past hundred years or so, especially in a low-flow condition or during drought years." Nevertheless, maintaining the hydro plant's electricity production was an important part of keeping the original water rights intact. Consequently, after a good deal of research and consultation, the early stages of the FERC licensing process for the existing (but unlicensed) hydroelectric project were initiated by the city in mid-2006.

In the meantime, since the hydroelectric plant was no longer powering the mining operations in Lead, the city needed to make a new connection with the local electric grid. In collaboration with the Black Hills Power Company, the city installed a new grid inter-tie system along with its associated switch gear and controls at the hydro plant. Remarkably, the original turbine/generators didn't need any repairs or upgrades, a testament to their rugged durability. The city also signed a power purchase agreement with Black Hills Power that provided for local electricity distribution to Spearfish, meeting part of the city's power needs from a local renewable resource. The plant generates approximately eighteen thousand MWh per year.

Even though the run-of-river hydro project had been in operation for almost a hundred years, the federal regulatory and approval process proved to be a challenge. "We started the initial consultation back around June 2006, but we did not get the FERC license issued until April 2011," Johnson says. "Because we are a small community with limited expertise, we relied very heavily on our consultants as well as our legal representation to walk us through the process. Nevertheless, a lot of people in town followed the process pretty closely because Spearfish Creek is the focal point of our community as well as our namesake, so it's a very sensitive topic in the community and people here are very possessive of it." The fact that the lower portion of Spearfish Creek that runs through the community is a blue-ribbon trout stream was an added incentive to protect the water quality and fish populations.

Because virtually all of the land occupied by the hydro project (fifty-seven acres) is located within the Black Hills National Forest, the USDA Forest Service played a major role in the process. In addition, the South Dakota Department of Environment and Natural Resources took an early interest as well. "The DNR was totally supportive of our application from the beginning because of the water rights," Johnson says. The city received a Section 401 Water Quality Certification from the state DNR in November 2010. However, there was a lot of local frustration with the role played by the Forest Service. "The frustration was with the Forest Service's ability to have mandatory conditioning authority, and a lot of our time and effort was spent trying to get the Forest Service to cooperate." Originally, there were eighteen Forest Service conditions placed on the project, but that number was significantly reduced in the final FERC license.

But there was another problem. The state DNR and the Forest Service ended up disagreeing over who had the authority over some of the conditions being imposed on the project, leaving the city stuck in the middle. FERC remained neutral. "We felt like we were the guinea pig. We really didn't want to get involved in this issue, we just wanted the license," says Johnson. "In the end we were able to get a compromise, but it literally took getting the two agencies into the same room in order to work it out." The city ultimately agreed to a long list of environmental measures designed to maintain appropriate water flows at various times of the year, prepare water-quality progress reports, help to relocate fish trapped in the project's forebay, install nest boxes for the American dipper (a bird that feeds on small organisms in the creek), and implement a Historic Properties Management Plan, among other provisions.

Throughout the process, the city had the strong support of the community. "The support for the project was huge, and it still is," Johnson says. Spearfish has already benefited from the hydro project, and will continue to do so for many years to come. "The sale of electricity helped us pay for the plant, it's paid for all of our licensing costs as we've moved throughout the project, and it continues to generate excess funds above and beyond the operating costs," she continues. The city's long-term plan is to use the funds for more recreational enhancements. "We've purchased additional creekfront property to try to provide more park space because most of our park system follows the Spearfish corridor that runs through our community. Those are the kinds of projects that we will use the additional funds for in the future for the benefit of the community."

Asked what advice she would give to other communities considering a similar project, Johnson responds, "Absolutely do it. Especially if it doesn't require a dam. The gains that you will receive from it, both from a renewable power supply standpoint and whatever other benefits are derived from it, are worth the effort. Just be sure that your consultants are good; with good consultants your chances of being successful are so much better."

Because water is relatively scarce in many of the western states, Johnson acknowledges that the opportunities for other river-based community hydro projects can be somewhat limited in the region, but she sees more options. "I have a background in wastewater treatment, and as I travel around I look at potential resources that we should be taking advantage of—like discharge pipes and the potential to produce hydroelectricity from them. I just think there are so many options out there, especially with the new micro-hydro technology that's available now; I think it's crazy not to be taking advantage of that for electricity." A growing number of communities nationwide would seem to agree.

Despite the nearly five years of hard work, Johnson has no regrets about her involvement in the Spearfish hydro licensing process. "I've had several people ask me if I had it to do all over again, would I still do the same thing? My answer is 'Yes, without a second thought.' It's definitely been worth it in the end."

Hamilton, Ohio

It has been estimated that of the eighty-two thousand dams in the United States, only 3 percent are used to generate electricity, offering a lot of potential for additional hydro generation. The prospect of constructing a new hydropower project without needing to build a dam has been increasingly attractive for private and municipal investors, and the number of proposals for new hydropower projects in this country has risen by about 30 percent in recent years, according to FERC officials. Many of these proposals are at existing federal dams, and one of the most ambitious of these is the Meldahl Project.

The City of Hamilton, Ohio, and its partner, American Municipal Power, Inc. (AMP), together hold a fifty-year license issued by FERC in June 2008 to develop a 105-megawatt hydroelectric generating facility called the Meldahl Hydroelectric Project, on the Ohio River at Foster, Kentucky, near Augusta, approximately forty miles east of Cincinnati. The City of Augusta previously held a 1995 FERC license for a similarly sized project (later reduced to 68.7 megawatts) at Meldahl, but lost it in 2005 because construction had not yet begun despite ten years of deadline extensions. The City of Hamilton, which had unsuccessfully applied for a project license in the 1990s, saw an opportunity to try again and filed for a new FERC license in October 2006. "We had been pursuing a license since around 1990, so we made it very clear that we were interested," says Mark Brandenburger, director of special utilities projects for Hamilton.[8] The Hamilton application was accepted, and the subsequent FERC licensing approval process took about three and a half years, relatively fast for a project of this size.[9]

The Meldahl Project includes a 105-megawatt run-of-river hydroelectric plant now under construction at the Captain Anthony B. Meldahl Dam and Locks operated by the U.S. Army Corps of Engineers. At a construction cost of more than $504 million, the hydroelectric turbine powerhouse is being built on the Kentucky side of the river because the locks, used by river traffic, are situated on the Ohio side. The Meldahl Project's proposed 138-kV transmission line will carry its electricity to the regional electric grid by crossing the Ohio River and running about two miles to an existing interconnect in Clermont County, Ohio. Construction began in May 2010. Once it becomes operational (scheduled for 2014), Meldahl will become the largest hydroelectric power plant on the Ohio River.

The new hydroelectric facility will occupy about eighty-one acres of federal lands on the south (Kentucky) bank of the Ohio River. The project will include an 1,850-foot-long intake channel, a large powerhouse added to the existing dam that will contain three horizontal bulb-type Kaplan turbine/generators[10] with a total installed capacity of 105 megawatts, and a tailrace channel directing water back into the river. The turbines will have

FIGURE 8-3. Aerial view of the hydroelectric powerhouse under construction at the Meldahl Dam and Locks on the Ohio River. Photo by Aerial Innovations of TN, Inc.

the capacity to generate 489 gigawatt-hours of electricity annually. The U.S. Army Corps of Engineers will control the low-head (twenty-four feet) flows available for the operation of the Meldahl Project. Operations will be coordinated with the corps's operation of the locks and floodgates to maintain the navigational area above the dam.

In addition to a wide range of environmental provisions, the FERC license also calls for enhanced recreational facilities and access adjacent to the project area, including parking areas, restrooms, foot trails, and other improvements for fishing, sightseeing, picnicking, boating, swimming, sunbathing, and camping. The plan specifically calls for fish structures, including a fishing access pier at Big Snag Creek sandbar, designed to benefit the recreational fishery downstream from the Meldahl Dam. This area was a popular public recreational site before project construction began.[11]

In addition to Hamilton, approximately forty-seven other cities through their membership in AMP are participating in the Meldahl Project, and AMP is providing the financing for all of them. "The financing is being done through the typical investment banking process, but it also incorporates some of the federal financing opportunities like Build America Bonds and Clean Renewable Energy Bonds," Brandenburger says. "AMP has a very large finance staff that devotes itself almost exclusively to financing." AMP is also building three other hydroelectric projects on the Ohio River at the same time as Meldahl, keeping the financing staff—and a lot of area contractors—very busy. Hydroelectric plant construction at Meldahl will peak with more than four hundred construction workers representing a wide variety of construction trades.

Hamilton's participation in the project was not always a sure thing. In 1999, the city council insisted that Hamilton's utilities become competitive—or they should be sold. A number of times after that, there were opportunities to follow through on the threat, but it never happened. "This project was our number one goal on the council," says Don Ryan, former mayor and councilman. "In 2000 we said, 'Go for it.' We knew what this could do for the city. It's an incredible feat for the city."[12]

Hamilton has relied on hydroelectricity for part of its municipal power portfolio since 1893. In 1963, Hamilton bought the approximately two-megawatt hydroelectric plant that once powered one of Henry Ford's car factories. Hamilton has also owned the seventy-thousand-kilowatt Greenup Hydroelectric

Power Plant located in Greenup, Kentucky, since 1982. The city's positive experience with Greenup is what inspired its interest in the Meldahl Project. Hamilton's municipal electric system provides power to twenty-nine thousand residential, commercial, and industrial customers and is a founding member of its AMP partner. AMP, a nonprofit corporation founded in 1971, serves as a joint action agency for a consortium of electricity generating or distributing cities located in Ohio and several other states. Hamilton's entitlement is 51.4 percent of the Meldahl Project's electric output, with other AMP member communities being entitled to the remainder, representing a remarkable multicommunity collaboration on an extremely ambitious renewable energy project.

The licensing for the Meldahl Project was not easy. "Getting the license was a major undertaking, especially when you consider that Hamilton began in the 1990s and didn't finally get it until 2008," Brandenburger says. "The regulatory process is very challenging; this is not something you can do on your own. You need to retain environmental and other consultants as well as a legal team to lead you through the federal and state regulatory requirements as well as the various permit requirements. You have to comply with federal law on water quality. You have to be careful what you are doing to the existing dam and lock so you don't damage it. You have to preserve the pool of water that constitutes the basis for navigation in the Ohio River. And you are dealing with multiple federal agencies who have interacting and sometimes overlapping responsibilities."

Kent Carson, the senior director of communications at AMP, agrees with Brandenburger. "One of the things that we hear over and over again is that it's extremely difficult for a community to take on a project by themselves," he says. "Obviously Hamilton did it in the 1980s with Greenup, but it's become increasingly difficult in recent years because of the risks, especially the construction and financial risks. What has made these Ohio River projects attractive to our members is that the risk is spread out. Secondly, they don't have to take all of the hydropower production. Diversification of generation portfolios is very important to public power entities, so if communities can band together it certainly makes these projects more attractive."[13] AMP has wind, solar, and landfill gas in its renewable energy portfolio (along with coal and natural gas in its fossil fuel portion), but generally feels that hydropower is the best choice overall as a renewable resource, according to Carson.

Not surprisingly, there has been a lot of community support for the Meldahl Project. "Hamilton's support for this project is very strong," says Brandenburger. "I think they take a great deal of pride in having a voice in the decision-making process. That's true for all of the communities that are involved through AMP as well. I have attended the board meetings at AMP and I can tell you that each of the communities that sits on that board is very interested in these Ohio River projects."

The Ohio River dams currently being retrofitted for hydropower by AMP were originally built by the U.S. Army Corps of Engineers to help maintain the channels for interstate commerce and to provide some flood control as well. "For forty years or so, these dams have been on the river and represent a largely untapped resource because they can be renovated to generate electricity," Carson says. "The river is there, the dams are there, and the water is flowing twenty-four hours of every day, so why wouldn't we make use of that resource?"

"The dams were built by the Corps of Engineers in the 1960s with the idea in mind that hydro plants would be added to them eventually," Brandenburger adds. "If you look at the Corp's drawings from the '60s they all show space for a hydroelectric plant. They knew that the time would come when it would be economically feasible, and that time is now."

Biomass

Unlike fuel oil, propane, and natural gas,
biomass has a history of stable prices unaffected
by global economics and political events.
—BIOMASS ENERGY RESOURCE CENTER (BERC)

In November 2008, I attended the Vermont Grass Energy Symposium held at Shelburne Farms in Shelburne, Vermont. No, it wasn't a gathering of people trying to legalize marijuana. Rather, it was a fascinating day of presentations, discussions, and demonstrations of the potential for the growing, harvesting, processing, pelletizing, and burning of grasses as a renewable biomass fuel. One of the most memorable parts of the day was a live outdoor demonstration of using locally produced grass pellets for fuel in a special biomass pellet burner. The extremely hot and clean flame produced by the burner felt good on the rather chilly afternoon of the demonstration. While there are still a number of technical problems associated with grass pellets that need to be solved (especially mineral-related corrosion of boilers), the demonstration (and the entire symposium) suggested that grass-based biomass fuel offers a lot of potential. In the near future, you may not necessarily need to have ready access to vast areas of forestland in order to consider the use of biomass.

Of course, communities that are surrounded by forests have a great opportunity to become more energy self-reliant and to keep more of their energy dollars circulating in the local economy. In addition, since the price of wood-based fuels has remained relatively stable over time compared with fossil fuels, communities that switch to woody biomass (usually pellets or wood chips) can achieve greater control over their fuel costs while improving their energy resilience. Pellet-fired heating systems tend to be used in smaller buildings where the heating demands are not too large and space for storing and handling the pellets is limited. Wood-chip-fired systems, on the other hand, are generally better suited for larger buildings (or groups of buildings) where the chips can be delivered, stored, and handled with the aid of large automated systems.

As with liquid biofuels, when considering biomass it's particularly important to consider the energy return on energy invested (EROEI) in the resource. The fuels (likely diesel and gasoline) used to cultivate, harvest, process, and transport the biomass feedstock diminish its EROEI. Also important is the impact that new demand for biomass feedstock may have on regional farmland and ecosystems; biomass plant projects across the country have been opposed out of concerns that they will create incentives to overharvest forests or plant feedstock instead of food. But as with any renewable energy system, scale and context matter: A technology that might be destructive at utility scale can be quite sustainable and beneficial at a small, local scale if organized and operated properly.

In this chapter, I describe three examples of local biomass projects: a relatively simple but very successful pellet-fired heating system for a small rural hospital in Oregon; a larger, more complicated community district heating system that relies on mill waste for fuel in British Columbia; and a much larger chip-fired, combined heat and power (CHP) project for a private Vermont college.

Harney District Hospital, Burns, Oregon

Harney County, Oregon, located east of the Cascade Range, is larger than New Hampshire in area but has a population of only around seventy-five hundred. The diverse landscape includes high desert countryside, sagebrush, lakes, areas of ponderosa pine, a wood products industry in decline, and high unemployment. The Harney District Hospital located in Burns is the county's largest employer.

In 2005, the hospital's board was developing plans for a new hospital building to replace its 1940s-era facility. After considering a number of options, Jim Bishop, the hospital's CEO, wanted to include a wood pellet heating system in the plans. The architect and the general contractor didn't like the idea. Bishop, who had been an international wood products consultant before coming to the hospital, was convinced that the pellet system would not only work, but also save the hospital a lot of money. "Their reasoning was that they had never done it before, and they didn't know anybody else who had done it before either, so they were just extremely uncomfortable with the idea. They felt the pellet system was going to be a high-risk project," Bishop recalls.[1]

Bishop admits that there *was* a certain amount of risk since there was no U.S. distributor at the time for the Austrian Köb boiler he had in mind. But he had seen firsthand how routinely pellet boilers were used in Europe, and just how efficient they were. "They have a very advanced technology in pellet boilers there," he says. Fortunately for Bishop, several members of the hospital board had also worked in the wood products industry and had no trouble understanding the potential benefits that a wood pellet system could offer. They felt that an automated pellet boiler was a good match for the hospital staff, who wanted to devote their time and energy to running a hospital, not a heating system. With strong board support, Bishop was finally able to convince the architect and the general contractor that the hospital wanted to move forward with the pellet system. "Once they realized we were really serious, they designed an absolutely wonderful system for us," Bishop says.

The system includes a pellet-fired, half-million-Btu-per-hour hot water boiler located in the parking lot, connected to the hospital's hot water heating system by underground pipes. The hot water system is then connected to sixty-four water-source heat pumps, which control the sixty-four zones in the fifty-five-hundred-square-foot building. The pellets are stored in a fifteen-foot high, thirty-ton-capacity grain storage bin located next to the boiler. The entire system only takes up four parking spaces.

"We bought the pellet boiler assembled, plumbed, and wired; it was transported in a twenty-foot shipping container, along with the grain bin and the chimney in a second container," Bishop explains. "The boiler installation was simple. It took two guys two days to install the whole thing onto a pre-poured concrete pad. The cost for the boiler was just under $150,000, delivered from Austria." The total cost for the entire system was $269,000, including the boiler, the prefabricated building, all the mechanical systems, fees, and permits.[2] The system, financed with a USDA loan, Medicare reimbursements, and a local bank loan, was purchased from a dealer in Vancouver, British Columbia, the nearest one at the time that sold Köb boilers. The system now runs seven days a week, year-round, providing hot water in the summer, and space heating and hot water during the winter. Hospital staff check the pellet

FIGURE 9-1. Pellet-fired boiler and storage bin at the Harney District Hospital in Burns, Oregon. Photo by Harney District Hospital.

system controls daily by computer. Annual fuel costs are about $9,000 for forty-five tons of residential-grade pellets.

Ever since it was installed in 2007, the pellet heating system has worked even better than Bishop had hoped. "I think we had to replace one circulating pump back in June of 2007, but otherwise it has just worked beautifully," he says. "We shut the system down every two or three months and clean the boiler and dump the ash. That's it." The small amount of ash produced (about thirty gallons) is given to local gardeners. From a financial standpoint, the pellet system has done even better. "It's far exceeded our goals," Bishop continues. "Oil prices went through the roof after we started operating the system. We originally projected a payback in fifteen to eighteen years, but we figured that we got a three-year payback on this project because of the high oil prices. It was way, way better than we thought, with much less maintenance than we had anticipated because it's so automated."

Bishop is also pleased that his original decision to rely on a local renewable energy resource has proven to be a good one. "We live in Oregon, we are surrounded by wood. We all think it's a good thing to grow trees and use them, and it's just crazy to try and bring in some other energy source that we have far less control over," he says. "We made the assumption that wood prices would be stable for a long time, and so far that's proven to be a valid assumption. We were paying $200 a ton delivered back in June 2007, and today we are still paying $200 a ton."

However, there were some challenges setting up bulk pellet delivery to the hospital back in 2007. "Nobody was shipping pellets in bulk, it was all in forty-pound bags. So we worked with the mills, and ended up shipping the pellets in a grain delivery truck. Figuring out the logistics of that and making it happen was probably the second most difficult part of this after convincing people that the project made sense in the first place." The mill was located several hundred miles away; it took the grain truck well over five hours to get from the pellet mill to the hospital for the original delivery. Recently, a new pellet mill has opened just seventy miles north of Burns, which should cut the drive time (and the delivery charge).

Because the hospital's pellet heating system was one of the first in the region, it has attracted quite a lot of attention. "Since we put the system in, we've gotten visits from lots of people from all over the country who are now thinking about it," Bishop says. "A lot of them assume that it's going to be really exotic, and big and noisy, but it's not any of those things. It's almost totally silent, it's relatively small, and works like a dream." Following the hospital's lead, the local high school also converted their oil boiler to pellets, while the town where the new pellet mill was just built converted several of its municipal buildings to pellets as well. Schools in other area towns have also converted.

"I'm really proud of it," Bishop says of the hospital's pellet heating system. "Of course, it's always nice to be right. Back then, there were so many people saying that it was risky or dangerous. But now we've got this beautiful, brand-new hospital in a tiny rural town, and we've got one of the most energy-efficient and carbon-neutral systems that there is. We get visitors from Eugene and Portland, and it just kind of blows them away that we are so far ahead in some things. They just can't believe that we are so green and energy-conscious. And we say, 'Yes, but we also saved a ton of money.'"

Bishop has a number of suggestions for others who might be thinking about installing a pellet-fired heating system. "My advice would be to not necessarily listen to the people who try to discourage you," he says. "There are a lot of people who have fixated on a certain technology and a certain way of doing things in heating and cooling that kind of puts them in a box. I would at least take the time to visit a place like ours and just look at what we have. Bring the architect or engineer that would design the system and take a look. Once people see it, they get it. These days we're not the only one anymore; there are a lot of other pellet systems being installed so there is a lot of good experience with this technology. Now there are U.S. distributors of the equipment that we are using, so that's not an issue anymore either.

"We blazed the trail for a lot of people, and it's much easier now," Bishop concludes. "Just be open to other possibilities, and also think long term. I've talked to too many people who say, 'Well, oil prices today are $88 a barrel, so we are going to base our planning on that price.' But those energy prices are highly volatile. I would prefer to bet on something that seems to have less volatility and is more local."

Revelstoke, British Columbia, Canada

Located in the scenic and forested Columbia River Valley of British Columbia, the small city of Revelstoke (population eight thousand) has been described as a mountain paradise. This is no surprise, as the city is blessed with abundant natural resources along with stunning views. With two national parks at its doorstep and a fast-growing ski resort with the biggest vertical drop in North America, Revelstoke offers visitors numerous adventure opportunities, including skiing, hiking, mountain biking, fishing, whitewater rafting, canoeing, and more. What most visitors might not notice is that Revelstoke also offers a great example of how a small community can make use of one of its abundant local resources to increase its energy resilience—and save money over the long term.

In the late 1990s, the city's largest employer, the Downie Mill, was faced with potential closure because of air pollution from its old "beehive" waste wood and sawdust incinerator. Local community members saw an opportunity to use the waste wood for co-generation of heat and electricity—specifically, to fuel a district heating system for the city and to produce electricity to sell to BC Hydro, the provincial utility. In 1997, Natural Resources Canada funded the first of a series of studies to review the potential for these ideas. At the time, the low price for electricity and high capital costs made the co-generation idea unattractive, although the district heating proposal made sense on its own: It would improve local air quality, increase the city's energy self-sufficiency, create local jobs, and keep energy dollars circulating in the local economy while reducing greenhouse gases. It was also less expensive than the CHP proposal.

"We wanted to do a combined heat and power project that used all of the residual wood waste from the mill, about seventy thousand tons a year," explains Geoff Battersby, a retired physician, former Revelstoke mayor, and energy project coordinator. "It would have generated about 4.5 megawatts of electricity, but we only had a viable market for about 20 percent of the residual heat. The projected cost came in at $18.5 million, and based on 4.5 megawatts of generation that meant we would be spending around $4 million per megawatt. It just wasn't economically viable, so we switched to a 1.5-megawatt heat-only project, which ended up being an approximately $6 million project."[3] This included $3 million for the central plant, $2 million for phases one, two, and three of the district heating system (including design and engineering), $1.1 million for energy transfer stations, and the remainder for construction financing and other costs. While this was still expensive on a per-megawatt basis, it was an investment that appeared to be manageable, and the city saw other values than just the bottom line.

Initially, the city had hoped that the private sector would play the primary role in the project, but that didn't happen. "Most of the interest from business was for contracting on the project engineering, but the companies that might have participated directly didn't see sufficient returns in the proposal for it to be attractive for them," Battersby says. Undaunted, the city moved ahead with the project on its own in collaboration with the mill.

In 2001, the Revelstoke Community Energy Corporation (RCEC), managed by a volunteer board, was formed as a wholly owned subsidiary of the city to own the project. Project ownership was modeled on a highly successful community/forest industry partnership formed in 1993 to own and operate a timber forest license through the Revelstoke Community Forest Corporation (RCFC). The city-owned tree farm produced an annual harvest of 3.5 million cubic feet of wood per year and helped Revelstoke become a major timber processing center for the region. The use of this city-owned, stand-alone corporation concept was adopted to spare the city any liability that might occur from RCEC operations.

The Downie Mill provided the one-acre site and free fuel, and shared the labor costs; it also signed a twenty-year contract to purchase energy from RCEC. About 50 percent of the energy is sold back to the mill in the form of steam to heat their drying kilns. The rest, in the form of hot water, is distributed to customers in the city via 1.4 trench miles of underground piping that makes up the district heating system. In phase one, the system heated the city's winter sports arena and aquatic center, community center, the high school, a local hotel, and a bed-and-breakfast. In phases two and three, between 2009 and 2010, additional municipal and private buildings as well as a church were added to the system and the total number of buildings served increased to ten. Revelstoke's district heating system was the first wood-residue-fueled system of its kind in British Columbia. It was also the first to be developed, owned, and operated by a community-owned corporation.

The project was not without its challenges. "It wasn't difficult dealing with the community at all; everyone was extremely supportive of the project," says Battersby. "The hard part was getting the money to do it." This proved to be a lengthy and frustrating process that required a lot of persistence. In the end, however, the project was financed through a Federation of Canadian Municipalities Green Municipal Fund grant and other grants (31 percent); various loans, including a credit union loan (33 percent); and from the city itself (through the purchase of preferred shares with a 7 percent interest rate), in part from the Revelstoke Community Forest Corporation (35 percent).[4]

The challenges were not restricted to financing, however. "We certainly had our share of start-up problems," Battersby recalls. "When we came online in June 2005, we expected to do phase two of our building connections in the next year, but it wasn't until 2009 that phase two was done." One of the main problems was that provincial regulations called for a twenty-four-hour operator presence at steam boilers above a certain size—and the project's 1.5-megawatt boiler was subject to that regulation (in Scandinavia, biomass systems up to twenty megawatts are routinely remote-operated, according to Battersby).

In a move to circumvent the twenty-four-hour requirement, Revelstoke installed a boiler that heats "thermal oil," which can carry more heat than can water. The heated oil passes through heat-exchange steam generators, to produce steam for the mill and hot water for the district heating system. Unfortunately, this somewhat complicated design also produced a series of expensive headaches for the system. "The thermal oil design allowed us to avoid the twenty-four-hour staffing requirement, but it added $400,000 to the initial system cost and caused unexpected oil contamination in the hot water," Battersby says. "This was just another problem that developed, but not something that one would have expected might happen, and it was very difficult to identify. The tubes in a steam generator have now been replaced three times because of corrosion due to oxygen in the water, costing us $40,000 each time."

In addition, there were feed problems with the stringy cedar bark that was part of the original waste fuel from the mill. Eventually, the boiler plant had to switch to straight sawdust. "Sawdust is a nice premium fuel because it's consistent in size. Twenty-five percent of the output from the mill is sawdust, but we're still only using 10 percent of the mill's total residual output." Then, in December 2009, there was a fire in the plant. "It probably was electrical in origin with a motor, but they never did determine what the cause was," Battersby says. "We had to use propane in the backup boiler for six weeks while we were repairing the damage, and that really affected the bottom line."

Nevertheless, at this point, the bugs seem to have been worked out and the project has met most of its original goals. "Number one was to get rid of the beehive burner at the mill," Battersby continues. "We can't claim that we were solely responsible for its removal, but it's gone nevertheless. We also have cleaner air. When the beehive was still being used, a lot of fly ash covered the neighborhoods that surrounded it. So that's gone too. Also, greenhouse gases have been reduced. We are currently displacing around thirty-five hundred tons of greenhouse gases per year by virtue of displaced propane usage in the community, since most of the buildings we are heating were previously heated with propane. It has reduced imports of propane and kept the money formerly spent on it circulating within the community. We had also talked about the project being a non-tax financial resource for the city, but that has not been the case so far. We are six years into it, and a significant portion of our debt will be gone at the end of year ten, so we'll be producing revenue for the city after that. In the meantime, it hasn't cost the city—it just hasn't generated the income we had expected."

In retrospect, it's just as well that Revelstoke's community energy project was downsized from its original 4.5-megawatt CHP design. With the downturn of economic activity after the Great Recession, the Downie Mill has cut back to about 60 percent of its previous output. "If we had built that larger co-generation plant we'd be in big trouble now, looking for more fuel and having the extra expense to get it here."

Battersby offers some advice to other communities that might be considering a similar project. "Number one, make sure that you have an adequate market for the energy that you want to produce," he says. "Although it did not make financial sense in our case, other communities should consider doing combined heat and power to try to get the maximum potential from the fuel. Hiring consultants that have expertise in doing CHP plants and district energy systems is key. You have to know about your fuel supply too. Because these projects are so capital-intensive, long-term financing is required, and you need to have a long-term, assured supply of fuel. And you have to know what the long-term costs are going to be as well. We expect that when the first twenty years are up, we will spend at least $30 per wet ton[5] for fuel, and maybe more."

Battersby also notes that economies of scale are hard to achieve when it comes to small plants. He recommends keeping the design as simple as possible. "I would certainly avoid producing both steam and hot water, and just keep the hot water part. In the past, steam was often used for dry kiln operations, but nowadays kilns run on hot water. Another factor in a small community like ours is finding people with the capacity to run a plant like this. You don't need a power engineer, but it's important to at least have someone with experience working with high temperatures."

Battersby acknowledges the positive role played by the Downie Mill. "We're fortunate to have really good relations with this mill," he says. "I doubt that we could have found another corporate citizen that would have been so helpful. They really wanted to get rid of that old waste burner, so they were very accommodating. There was no market for their mill residue when we signed our agreement with them, but within a year there was, and they could have had a price for it."

Despite the problems the project has experienced, there is a good deal of community pride associated with it, according to Battersby. "It's a real flagship property," he concludes. "We've gotten all kinds of accolades and have won several awards for having done it. We have people who want to know about it all the time, and we could be away making speeches about it every month if we wanted to. We are really proud of it."

Middlebury College, Middlebury, Vermont

In 2004, Middlebury College, located in Middlebury, Vermont, committed to cutting its carbon emissions 8 percent below 1990 levels by 2012. At the time, the college burned about two million gallons of No. 6 fuel oil[6] every year to fire its boilers that heat the campus through a district heating system. Replacing that oil, or at least half of it, with a sustainable local alternative—biomass wood chips—soon became the focus of attention. In 2006, the college approved a proposal for a biomass plant, while students proposed a new, more aggressive goal of carbon neutrality by the fall of 2016—a goal that included the biomass plant and additional energy saving measures. The following year, the college trustees adopted the carbon neutrality proposal, and construction on the $11 million combined heat and power biomass gasification plant began. In 2008, the 8.8-megawatt biomass-fueled power plant was completed, and in early 2009 the CHP plant began full operation. The new facility was connected to the existing heating infrastructure and did not require any additional operating staff.[7]

The role the students played in the project was significant, according to Tom Corbin, the director of business services at the college. "The students were the big driver behind the carbon neutrality pledge that the trustees ultimately passed," he says. "First, they laid out the challenge. Then, we set up a Carbon Reduction Initiative Working Group—comprised of faculty, staff, and students—that got together to come up with ways that we could reduce our carbon footprint." The focus quickly turned to space heating, the single largest part of the college's carbon footprint, and a number of environmental studies classes began to look into what alternative fuels might be available. Biomass soon rose to the top of the list. "The students were involved in the initial thrust to do this, and then in coming up with the alternatives, and then studying whether or not those alternatives would work on a sustainable basis," Corbin adds.[8]

FIGURE 9-2. Biomass-fueled combined heat and power plant at Middlebury College in Middlebury, Vermont. Photo by Brett Simison.

As a nonprofit, private educational institution, the college was not eligible to receive the federal tax credits normally associated with this type of renewable energy project. Nevertheless, interest in the project remained high. "The real incentive for us was the price of oil, our inability to control it, and the fact that we were subject to an extremely volatile price for fuel," Corbin says. "There were also questions about the possibility of future interruptions in the supply of oil. The fact that we were sending all of this money out of the county, the state, or even the country was another concern. So, when we switched over to biomass, we gained greater control over our energy future, and the money is now being spent more locally; that's important to us."

In the biomass gasification system employed at the college, wood chips are superheated in a low-oxygen chamber, where they smolder and emit wood gas. Oxygen is then introduced on the backside of the boiler, causing the gas to ignite, producing heat (at temperatures of more than eleven hundred degrees Fahrenheit) to make steam that is distributed throughout campus in underground pipes for heating, cooling, hot water, and cooking. Exhaust from this process circulates through a separator, forcing larger particles to drop out. The exhaust then enters what is known as a bag house, where it passes through a series of filters that remove fine particulate matter. The filtration system is rated to remove 99.7 percent of particulates, so most of what is seen coming from the smokestack is water vapor. In addition, this system uses the exhaust from gasification to preheat the water entering the boiler and excess steam pressure to spin turbines that co-generate electricity before the steam is distributed throughout the campus. The co-generation produces three to five million kilowatt-hours of electricity per year (about one-fifth of the campus's electricity needs), all of which is used by the college. Additionally, the ash produced by the combustion process is used by local farms as a soil amendment.

The biomass plant was projected to consume around twenty thousand tons of wood chips per year, displacing half of the heating plant's annual two-million-gallon fuel oil consumption. During the first year of operation, however, things got off to a slow start. "It took longer than anticipated to learn how to source, handle, and gasify the solid fuel," says Corbin. "We had some fuel issues in the beginning. We found out that, while in theory you can vary the type of chip by the truckload, in practice it's not a good idea. We also got some dirty loads, and found out that you can't use dirty chips, so we really had to focus on getting a cleaner chip than what we started with. Once we got the clean hardwood chips coming in, we began achieving the target, so now the plant is operating well. We're using a million gallons less of No. 6 heating oil now."

When the college first began considering biomass in 2004, the price of No. 6 fuel oil was 89 cents per gallon. By the summer of 2008, the price had risen to over $3 a gallon. At that price, the annual savings resulting from the switch from oil to wood chips would have amounted to $2 million annually. The price of heating oil has since retreated somewhat, but the college is still saving about $1 million a year, according to Corbin.

One key part of the plant's design is its fuel flexibility. "The way the plant is constructed, we could switch to a different solid fuel if necessary," Corbin says. "So we could go to a compressed grass cube or something similar. If we wanted to switch to wood pellets we would have to put up a storage silo, but that's not all that difficult. This gives us the flexibility to switch fuel in the future if the solid fuel market changes. If someone else were to do this, they wouldn't want their facility to be so unique that they would be stuck with just one fuel that they couldn't get away from."

As projected, the biomass plant has cut the volume of greenhouse gases emitted by almost 12,500 tons per year, or about 40 percent. The facility receives its supply of wood chips from within a seventy-five-mile radius of Middlebury to keep it as local as possible and to minimize transport expenses. As part of this strategy, the college has been conducting experiments with growing its own fuel supply on ten acres of local agricultural land. In this collaboration with State University of New York's College of Environmental Science and Forestry in Syracuse, Middlebury College used fast-growing willow shrubs on its own land as a locally produced fuel. If the project succeeds, local farmers could grow willows on their marginal land and chip it to generate additional income for them and reduce the need to haul chips in from farther away. However, the willow experiment has run into a few problems.

"The willow planting went well, the interim cultivation was fine, and the harvest went well enough even though the ground was a bit wet," Corbin explains. "But the gasifier didn't like it at all; it actually put the fire out. The problem with a gasifier is that it doesn't react as quickly as a woodstove. When you add something new to the fuel, you don't see an instant reaction. Unfortunately, it's also slower to react when you try to make an adjustment, so by the time you make the adjustment, it's already too late."

The college also learned a lot about willow harvesting. "We found out in the process of harvesting that you get more dust because it's going through a corn chopper, so it needs to be screened before you can use it. We also learned that if we decide to plant another crop, we would need to make it easier to harvest. What nobody really understood until we started harvesting was that you leave about a four-to-six-inch stub, which will go right through a tractor tire—and tires on harvesters are expensive. So the rows needed to be set a little bit farther apart, and we needed to leave a bit more room at the end of the rows to allow for easier turning. It was an experiment, and if we do go forward with it, we certainly know a lot more now than we did five years ago."

But that wasn't the only problem. In May 2011, a fire started spontaneously in the biomass plant's bag house filtration system. The bag house was damaged beyond repair and had to be replaced, causing a shutdown for the biomass plant that lasted three months. Investigators concluded that the fire had not been deliberately set, but were unable to determine what actually caused it. The new bag house design includes increased fire detection and suppression systems to minimize the possibility of a repetition.

Despite the problems with the bag house and the willows, college officials are pleased with the biomass plant and consider it to be a success. "We have saved a lot of money," Corbin says. "We are in a much better position to ride out some of these fluctuations in oil prices than we were. Also, we now have a fair amount of experience with this, and have raised the profile of biomass and shown that it can be done on a campus our size. We think others may follow. We certainly have had enough people tour through the plant in the last few years to show that there is a lot of interest."

Biogas

We're committed to biogas because we
believe it will benefit rural America.
People talk of the local foods movement;
well, this is really the local energy movement.
—DALLAS TONSAGER, UNDERSECRETARY FOR RURAL DEVELOPMENT, USDA

For centuries, alchemists dreamed of turning base metals like iron and lead into gold. They never succeeded. But what about turning base materials such as sewage, manure, and organic waste into electricity, heat, or vehicle fuel? This new alchemy, involving the use of biogas, is being successfully practiced in a growing number of communities across the nation, and offers much promise for greater local energy resilience.

One of the earliest and most successful farm-based methane initiatives is Central Vermont Public Service's (CVPS's) nationally recognized "CVPS Cow Power" manure-to-methane initiative, which began in 2005 with one farm and fifteen hundred dairy cattle. By 2010, the program produced approximately 1.4 megawatts annually from eight farms. CVPS customers can opt to get a percentage of their electricity from manure-to-methane projects, with participating farms being added each year.[1] Vermont is not the only state to host manure-to-methane projects. On-farm use of methane digesters was reported by 121 operations in twenty-nine states according to the U.S. Department of Agriculture in 2011.[2] One successful, locally owned start-up in Washington State that recycles local farm and food waste into renewable electricity leads off the case studies below.

There are a lot of other potential community biogas projects to be found at wastewater treatment plants. The United States already has a total of fifteen hundred wastewater-treatment-facility-based anaerobic digestion systems that produce biogas. But with only 250 of those plants making productive use of their biogas, there is obviously an opportunity for wider utilization of this local energy resource.[3] We'll take a look at one of the most ambitious municipal biogas initiatives in the nation in Des Moines, Iowa.

The use of landfill gas (LFG) for electricity generation is fairly common in the United States, where around 541 landfills are collecting it for energy—although it is also somewhat controversial. LFG can contain toxic contaminants, potentially creating a public health risk if combusted incorrectly; LFG-to-energy arrangements have created incentives to *increase* material going to landfills in some places; and some LFG capture setups inadvertently release "fugitive" methane into the atmosphere, erasing any greenhouse-gas-reduction benefit.[4]

Only a handful of municipalities are converting their landfill gas to compressed natural gas (CNG) or liquefied natural gas (LNG) for use as vehicle fuel. The main reason is that, until very recently, the biogas conditioning systems needed to clean up the landfill gas to meet vehicle engine fuel specifications have been very large and expensive. We'll look at a small, innovative project in Dane County, Wisconsin, that has solved that problem.

Farm Power Northwest, Washington

Farm Power Northwest LLC, a Skagit County, Washington, company that harvests methane from manure, was founded by brothers Kevin and Daryl Maas. The company's facilities are designed around anaerobic manure digesters that produce methane-rich biogas used to fuel generators that produce electricity. The by-products of the processed manure go back to the partner farmers as organic fertilizer and free bedding. Farm Power has two digester facilities in operation and a third under construction. "We were a true local start-up, with no big corporation behind us and no fancy marketing," Kevin Maas says. "It's just me and my brother and some local investors."

The brothers grew up in Skagit County and enjoyed the rural lifestyle that comes with living in western Washington. They believe that in order to maintain Puget Sound's balance between economic growth and outdoor quality of life, farming must be both economically profitable and environmentally sustainable. Manure digesters would seem to be an obvious strategy to help achieve both of those goals, but despite having nearly five hundred dairy farms Washington has only five digesters.

Farm Power's goal is to serve as many Pacific Northwest dairy farms as possible with manure digesters. Most dairy farms in western Washington and Oregon are too small to build their own digesters and successfully market the products. That's where Farm Power comes in: They work with groups of dairy farmers to design regional digesters serving multiple farms.[5]

The original inspiration for Farm Power came in 2004. While a high school teacher, Kevin toured the Vander Haak Dairy located near Lynden, home of the first methane digester in Washington. He assumed that all the farmers in the area would soon be building digesters of their own. But a year later, nothing had changed, and Kevin decided to pursue the idea himself. He enrolled at Bainbridge Graduate Institute and ultimately earned an MBA in sustainable business. His last project at the school was Farm Power's business plan. Daryl joined Kevin in 2007 to form Farm Power.

Their first digester, located in Rexville, took two and a half years from concept to start-up. The technology was relatively simple. The challenge was bringing the farmers, the bankers, the regulators, the environmentalists, and the utility together for the project. Two adjacent dairy farmers who together milk twelve hundred cows agreed to supply the manure, which they had been storing in open lagoons. "We located between the two dairy farms," says Maas. "We had known these guys for most of our lives, so that helped convince them that we knew what we were doing. A big part of the process is just convincing farmers to sign anything. But once we got the farmers to commit, we leased about three acres from one of them, and went out and applied for grants and loan guarantees. We found a sustainable bank that was willing to lend us most of the project money; we were also awarded a couple of grants and a loan guarantee, which made the bank happy."[6] Farm Power also sold methane emission reduction credits to the Climate Trust to help fund the project.[7] With all the funding and permits in place, a contractor was hired to build the digester.

The brothers selected one of the most popular and successful digester designs, made by GHD, Inc. of Chilton, Wisconsin, to ensure that their first project would run smoothly. "Over half of the digesters in the country are from GHD, so we felt they were our best choice," Maas says. "When we turned it on, it just started running. It's a good design, and we learned quite a

FIGURE 10-1. Farm Power's first digester, in Rexville, Washington. Photo by Farm Power.

lot about manure handling along the way." The project, with its metal building containing a large orange engine/generator connected to the concrete digester vessel by yellow pipes, came online in August 2009. In return for their manure, the two participating Rexville farms receive fertilizer and digested fiber from the facility, which they use for cow bedding—saving them about $100,000 a year because they don't have to purchase sawdust or straw.

Farm Power Rexville sells the electricity it generates to Puget Sound Energy's Green Power program, supported by the utility's customers who voluntarily pay a small premium on their electric bills to assist green energy projects. "When we signed the power purchase contract, it was one of the best arrangements in the country," Maas says. "We didn't have to provide any security or minimum delivery guarantees. If we produced power, they'd buy it, if we didn't, it wasn't a problem."

Farm Power now has two other projects in the ground: Farm Power Lynden, located near Lynden, and Farm Power Tillamook in northwestern Oregon (under construction). Each project is a wholly owned subsidiary of Farm Power. "All of our investors own a little piece of everything that we have done. That's attractive to a lot of them because it really reduces their risk," Maas says.

Between the Rexville and Lynden facilities, the two operating 750-killowatt generators produce enough electricity for about a thousand homes.

The Farm Power Lynden project, which came online in November 2010, uses manure from about two thousand dairy cows in its digester. In addition to generating electricity and producing free bedding for the participating farmers, it also utilizes the heat of the generator to produce hot water for heating a nearby greenhouse. The $3.9 million project received federal support with a combination of grants and loans; a $1.1 million portion came from a grant funded by the American Recovery and Reinvestment Act through the state department of commerce. Farm Power combined these dollars with money it got from the U.S. Department of Agriculture—$500,000 in grants and $2.4 million in loan guarantees.

"We were building right in the middle of the stimulus program, but we didn't go in expecting a Treasury grant or the investment tax credit because our bank will lend us money on construction if the project is designed well. Then, all of a sudden, the law changed and we saw right away that it was going to be amazing. So we took the ITC Treasury grant on the back end; it was not part of our original financing package. On the front end we received a USDA grant/loan combination, and that was really vital. We borrowed over half of the money for each project— so the half-million-dollar grants which the USDA gave us were really nice, but a loan guarantee for a $2.4 million loan is arguably even more important."

Rainier Biogas, LLC, Farm Power's fourth project, is planned for a group of farms nestled in the foothills of the Cascades near Enumclaw, Washington. The project had been in the organizing stage for almost ten years, but never quite got off the ground until Farm Power arrived on the scene. Rainier Biogas is a collaboration among three family dairy farms with a total of twelve hundred cows. The one-megawatt project has received incentives from a state department of commerce energy program grant and loan (again funded by federal stimulus dollars), the U.S. Department of Energy, King County, the USDA, and *Native*Energy. Vermont-based *Native*Energy sold carbon offsets that helped finance the project.

"For our follow-up projects we got some of this stimulus money that we didn't know existed in 2008, and that's helped us to grow faster than we otherwise would have," Maas says. But Farm Power has also relied on local investors. In the fall of 2010, using a Securities and Exchange Commission exemption, Farm Power raised around $750,000 in a private stock offering.

"We went back out to the community and were funded midwestern-style, by middle-class, mostly local investors," he adds. "It was really exciting."

The Farm Power projects have had their financial and technical challenges, but like many local renewable energy initiatives, dealing with inflexible permitting and regulation has been near the top of the list. "Once we got the farmers to sign commitments to a project, the biggest challenges have been with permitting," says Maas. "We had quite a struggle several years ago about whether we could mix other organic waste with the manure. The Washington State Department of Ecology said that if we did that, we would become a solid waste handler. We told them that the farmers would not be willing to have a solid waste handler on their farm due to all of the additional regulations, but they wouldn't budge. So we actually had to go change the law. It was crazy that we were not able to sit down with the department and figure out a reasonable compromise; we had to have the legislature do it for us. I think most of our challenges have to do with regulators who are unable to weigh all the pros and cons. We've heard that's a problem in other states as well." As a result of the law change, all of Farm Power's Washington digesters can now take in pre-consumer food processing waste along with the manure.

Despite all the hard work, the Farm Power projects have been a great success, according to Maas. "First and foremost, we define success when our farmers are happy, and secondly when our bankers are happy, and so far that's been the case with both groups. The farmers have also been surprisingly happy with our bedding product. It's worked a little better than we had expected. It hasn't always been perfect, but overall we're doing it right, and it's an outstanding product. The farmers get it for free and they're saving real money."

Maas is proud of what Farm Power has accomplished in just a few years. "I think these projects are great; I love to show them off to people," he says. "I think we have had a tremendous impact on the local farm community, and I'm glad we did it. It's not always been fun, but it's definitely been worth it."

Des Moines Metro Wastewater Reclamation Authority, Iowa

The Des Moines Metro Wastewater Reclamation Authority (WRA) in Des Moines, Iowa, has one of the most ambitious large-scale biogas operations in the country. The WRA serves a population of approximately five hundred thousand, treating thirty-six billion gallons of wastewater a year from sixteen area municipalities, counties, and sewer districts around Des Moines. The WRA is governed by a board of officials from Des Moines and the surrounding communities.

In 2009, the WRA generated more than 460 million cubic feet of biogas by co-digesting twenty-six million gallons of organic waste. The waste, produced by regional businesses, is trucked to the WRA's site, where it is added to the digesters. The digester complex contains six huge digesters, each with a capacity of 2.7 million gallons, for a combined total capacity of 16 million gallons. Biogas produced by the digesters fuels three 600-kilowatt co-generation units that produce over eight million kWh of electricity. The biogas also fires three dual-fuel (biogas or natural gas) boilers that produce building and process heat for the facility. About 40 percent of the biogas is sold to a nearby industrial facility owned by Cargill.[8]

FIGURE 10-2. Digester renovation work at the Wastewater Reclamation Facility digester complex in Des Moines, Iowa. Photo by Des Moines Metro Wastewater Reclamation Authority.

The $250 million Wastewater Reclamation Facility (WRF) owned by the WRA is more than a mile long and covers seventy-seven acres. It houses a series of treatment processes for the area's wastewater collection system, which operates around the clock and employs about a hundred engineers, managers, maintenance workers, laboratory technicians, and other support staff. The facility first came online in 1987.

The co-digesting started in 1990, when a local dairy was looking for a place to dispose of its whey waste, according to Larry Hare, the WRF's operations supervisor. "Initially, we started adding the whey to our digesters for a very small tipping fee just to see what the effects would be. We immediately saw what we expected, which was increased digester gas, and we became very interested in co-digestion. At that time I believe we just used one or two of the co-generators, and we wanted to make better use of them." Thanks to the expanded co-digestion activity, the WRF's three generators now run almost all the time, and there are preliminary plans to add two 1.4-megawatt co-generation units in the next few years.

The WRA initially focused mainly on local food processors for its co-digestion feedstocks, although it began to accept inputs such as pre-treatment waste from packing plants and restaurant greases. But with the rapid growth in the local biofuels sector, co-digestion waste soared after state officials began to crack down on the land application of wastewater and sludge from biodiesel and ethanol plants. "Here in Iowa, we have a lot of biofuels plants," Hare says. "We take a lot of the waste from just about all of the biodiesel plants and a few in surrounding states as well. We also take some waste from the ethanol plants, but not as much. The ethanol side tends to be 'hotter' [higher strength] and doesn't always work with our process, but the biodiesel side always has worked great for us." The more intense activity caused by the "hotter" feedstocks tends to give the digesters a case of indigestion, according to Hare. Annual revenues from the tipping fees paid by the waste haulers tend to fluctuate, but were around $1.9 million in 2009, and $2.4 million in 2010.[9]

The co-digestion initiative was extremely successful—and before long, it became too successful. The increased biogas production was initially absorbed by the co-generation units and dual-fuel boilers. But eventually they couldn't keep up, and a lot of excess biogas had to be burned off (flared) as there was no obvious market for it. Low natural gas prices further discouraged the

WRA from installing an additional process to clean the biogas and sell it to the natural gas utility. "But flaring the gas was just wasting it," Hare says.

At this point, the WRA began to look for an additional market for its excess biogas. It found it right next door at a Cargill plant that used natural gas and fuel oil to fire its steam boiler. The WRA approached Cargill with a proposal to sell some of its biogas to produce steam, and Cargill liked the idea. A six-hundred-foot pipe was constructed between the two facilities, and a pricing agreement worked out. "The project was about two years in the planning and construction phases," Hare remembers. "We began selling gas to Cargill in 2007. Since then, we have received $2 million from the sale of gas that previously was flared off. We still flare a little gas if we can't use it in-house or sell it to Cargill, but we try to keep it under 5 percent of biogas production. We have talked about other options like a compressed natural gas fueling station for vehicles."

In 2010, the WRA began a $21 million digester improvement project to upgrade its digester complex and biogas distribution system. The project, which is scheduled for completion in late 2012, includes replacement of the aging digester covers, the gas mixing system, and other technical improvements that should improve efficiency and reduce maintenance costs. The additional biogas production combined with new co-generation capacity should allow the wastewater treatment plant to move closer to net zero energy consumption, according to Hare.

While there have been a number of challenges along the way, one of the biggest was not technical but rather internal and cultural. "Probably the biggest challenge was just getting the mind-set of a major utility like ours to change and agree to try something new," says Hare. "With a lot of these things, we have been blazing a trail because they haven't been widely tried on this scale in the U.S. There is quite a lot of this sort of thing in Europe, but not in this country." Unlike many other local renewable energy initiatives, the regulatory and approval process for WRA projects has not been a major issue, according to Hare. "Luckily, here in Iowa it hasn't been bad. It's been mostly state approvals, and I would say we didn't have any major problems."

Hare offers some advice for others who might be considering a biogas project of their own. "They need to think about the scale of the system that they are putting in," he says. "We are kind of lucky; we have about fifteen million gallons to work with. That can absorb a 'hot' load with only small effects.

A lot of other systems that you see on farms or in communities are maybe hundred-thousand-gallon or ten-thousand-gallon digesters, and it doesn't take much to upset those smaller systems. If those systems are using something like dairy waste or manure, you really have to be careful when you are mixing in different types of waste. I think we've tried just about everything there is to mix in, from scrambled eggs to vinegar and much more. We've learned that you have to be careful about what you put in."

The WRA's co-digestion initiative has had multiple benefits. "It's turned out to be a savings of money and also of energy that was previously just being wasted. It's also turned out to be a revenue source for us that offsets some of our budget expenses." For many years, this has helped the WRA to maintain steady rates for area taxpayers, something they appreciate, according to Hare.

The facility has also become a draw for educational institutions. "We do a lot of tours with schools and colleges and other groups coming through. A lot of them didn't even know that this was happening, and have been surprised and pleased about it. Cargill, of course, is very happy about the project we've done with them. In fact, they have modeled it corporation-wide and shown what a public–private partnership can do."

Hare is proud of the co-digestion and co-generation projects, and is excited about the ongoing digester renovation work. "I think we have come a long way since we first started dabbling with whey back in 1990," he says. "When this upgrade project is complete it will allow us to continue to expand our biogas production, energy conservation, and our commitment to the environment."

Dane County, Wisconsin

The idea of fueling vehicles with garbage may sound a bit far-fetched, but that's exactly what the folks at the Rodefeld Landfill in Dane County, Wisconsin, are doing. Located in the south-central part of the state, the county is home to the state capital, Madison; the University of Wisconsin and Madison College; as well as more than sixty cities, villages, and towns.

Since 1997, the county landfill at Rodefeld had been using landfill gas (LFG) to produce about four megawatts of electricity. But the county was receiving a relatively low price from the utility for its LFG-generated electricity and was looking for a way to clean up the gas to pipeline quality so it could sell it as natural gas at a higher price. The county found that technologies to clean up the gas were expensive and not readily available for small projects like theirs. But then the price the county was paid by the utility for its LFG-generated electricity went up, while the price for natural gas went down. County officials abandoned the plans and continued to generate electricity, an activity that now generates about $3.5 million for the county every year. But the technology to clean the LFG presented another opportunity: producing biogas, in the form of compressed natural gas (CNG), to fuel county vehicles.

John Welch, the recycling manager/project manager for Dane County Public Works, and Mike DiMaggio, the solid waste manager, were both convinced that the BioCNG idea would work—if they could find the right technology. "We were looking at these systems and they were very expensive," DiMaggio says. "We wondered if someone made one that was small enough that you could just fuel a small fleet like ours, but almost nobody had one."[10] They turned to Cornerstone Environmental Group, LLC in Madison. Mark Torresani, a Cornerstone engineer, had already found that affordable small-scale technology to remove contaminants such as carbon dioxide, hydrogen sulfide, volatile organic compounds, and water vapor from the LFG was not available. But he thought that a cost-effective system could be developed for the project. "We worked on the engineering with a systems fabricator, Unison Solutions, LLC of Dubuque, Iowa, to try to bring the technology down to size," he says.[11]

The project eventually developed into a collaborative effort among Dane County and several public and private entities. Alliant Energy donated a trailer-mounted CNG fueling station originally built in the early 1980s. Team members included the Dane County Solid Waste Department, Cornerstone Environmental Group (design/permitting/site implementation), ANGI Energy Systems, Inc. (CNG fueling station refurbishment), Unison Solutions (gas conditioning system design and fabrication), and Madison College (CNG fueling station trailer repairs). After much work, the prototype system was installed in December 2010, and after some fine-tuning the first vehicle was fueled on March 18, 2011. "They came up with this system and it worked really well for a prototype. I was shocked," DiMaggio says. "I've been in this business for forty years, so I've been around a lot of R&D projects, and I was really surprised at how well it worked. We had a few minor problems, and a couple of leaks, but after some modifications it ran very well. It actually made our generators run better too." After cleaning and conditioning by the new system, the BioCNG meets industry and engine manufacturers' fuel specifications.

"The landfill gas coming into the conditioner is about 52 percent methane, and at the back end of the process we end up with about 90 percent methane. It's very pure gas," Welch says. "The carbon dioxide off-gas that's removed by the conditioning system goes back into our gas line and is used up in our electric generators. That's why co-locating this with a larger system that generates electricity makes so much sense; it's very complementary. From there, the biogas goes into the CNG fueling station, which can run on biogas like ours or conventional pipeline gas. It takes the gas, removes a little more moisture, and condenses it down to the three thousand or thirty-six hundred pounds per square inch (psi) that's needed in CNG vehicles. Then it stores it, and has a hose and nozzle to dispense it."

The patent-pending biogas conditioning system represents a significant breakthrough in a small-scale system that can economically produce small quantities of BioCNG to fuel modest local fleets of CNG vehicles. "What's unique about this project is that we are now dealing with quantities of gas that would have been deemed too small to do anything with in the past," notes Cornerstone's Torresani.

The initial system was designed to use a small amount of excess LFG measured at twenty standard cubic feet per minute (scfm), capable of producing about a hundred gallons of gasoline equivalent (GGE) BioCNG per day. "That can fuel around ten vehicles at our current usage," John Welch says. "The vehicles we are using do not travel a terribly long distance during the day, but we are starting to have more CNG vehicles for our parks and our highway department, which can go through a tank a day because they travel farther. In 2012 we should have twenty to twenty-five CNG vehicles." As a consequence, the BioCNG production system will be upgraded to a larger size in 2012 to handle fifty scfm of LFG, capable of producing about 250 gallons of GGE per day. With this larger size system, the BioCNG will cost about $1.78 per GGE. The system can be expanded further, if needed, reducing the cost per GGE to around $1. A local taxi company and a waste hauler have expressed interest in the fuel.

In addition to landfills, this technology can also be used in conjunction with municipal wastewater reclamation facilities as well as farm-based digesters. Along with the conditioning system size upgrade, the county also has money in its 2012 budget for a new fueling station. "We didn't want to go out and spend a lot of money on a new gas station until we knew how this was going to work," DiMaggio says.

The Rodefeld project faced a number of challenges. Trying to attract financing was at the top of the list. "We tried to find investors who were willing to invest in it, and we didn't have a lot of luck," says DiMaggio. "If Unison and Cornerstone had not stepped up to the plate, I'm not sure if it would have happened. They were certain that there was a market for this and wanted to go ahead and build it anyway."

Welch agrees. "The financing was a huge issue. The county board said they wanted to see one of these things run before they put any money into it. So, Cornerstone and Unison covered the cost of the first system because they had faith that they would eventually be able to sell more systems." The county was offered free use of the system for a year to test it out and decide whether they wanted to buy it. They did. The costs for the project going forward will be covered by municipal bonds and tipping fees, according to Welch.

But that wasn't the only issue. Because it was the first of its kind, permitting for the new system turned out to be another hurdle. Cornerstone worked with the local fire department and Wisconsin Department of Commerce, neither of which had ever permitted a similar facility. Cornerstone and Dane County staff worked with state and local officials to obtain the permits to handle combustible gas necessary to operate the system. The old fueling station, in particular, needed a number of modifications to bring it up to current safety standards and receive its permits.

Obviously, a landfill won't produce LFG forever, but this reality is factored into projects like this from the beginning. "The Rodefeld landfill is still open, and we probably have another ten years before it closes, and then another twenty-five to thirty years beyond that for usable gas production; so we are looking at around forty years," DiMaggio says. "Everything we built is skid-mounted, so that when we build the next landfill, we can just move the equipment over to the new one. The same is true with all of our generators."

The project has definitely achieved its original goals, according to DiMaggio. "Clean vehicle fuel, at a low cost—and it actually works. Besides, even though we have an active recycling program, landfills in general have been getting a bad rap for years, and this is a way to start to provide a better image."

Both DiMaggio and Welch think that BioCNG has a bright future. "In general, I feel that BioCNG *is* the future," DiMaggio says. "People talk about CNG and natural gas just being bridge fuels because they really are just fossil fuels, but I think that BioCNG is a huge step beyond that," Welch adds. "It really is something that is renewable because you can take biogas from wastewater treatment plants and farm digesters and use it to create a truly renewable fuel. I think that is where the future is going to be. On a life-cycle analysis, this is far cleaner than normal CNG, and you are looking at an 88 percent reduction of carbon emissions compared to gasoline."

Not surprisingly, there has been a tremendous amount of interest in the Rodefeld biogas-to-CNG facility since it opened, and Dane County and the project partners have shown the system to more than one hundred public and private groups and individuals. More than sixty people participated in a tour during the BioCycle conference in November 2011. "We do a lot of tours at the landfill, and the people are amazed that we can take gas from that hill and generate electricity," DiMaggio says. "But then we step out of the building and say, 'We've gone further than that, we are producing vehicle fuel.' Then they're even more amazed," Welch adds. "That overshadows all the electric generation we do when we show them that we are running our vehicles on garbage."

The Dane County BioCNG fuel project was the first of its kind in the nation and has served as a test of the small-scale technology's reliability. Following the success of the project, two additional projects, one at a landfill in St. Landry Parish, Louisiana, and another at the wastewater treatment plant in the City of Janesville, Wisconsin, are under construction. More are planned.

DiMaggio is enthusiastic about the BioCNG project. "I've worked in this business for forty years, and this is the most excited I've ever been about this," he says. "We were told that we couldn't do it, but we did. We're producing pipeline-quality BioCNG. Some people have asked, 'Why did you do this?' We did it because it earns us more money, but it also just makes sense to do it."

Liquid Biofuels

Nine years ago, I could jump into any Dumpster in the area.
They were overflowing, the lids were open, and the restaurateurs
were pleased for us to take as much as we wanted.
Now they're watertight, locked, and under contract.

—Lyle Estill

While all of the renewable energy sectors have had their challenges, the liquid biofuels sector—especially biodiesel—has faced probably some of the most serious in recent years. The ups and downs experienced by biodiesel producers large and small have been particularly gut-wrenching, and often related to inconsistent government policy. To be sure, the biodiesel sector has been subject to other forces as well. During the run-up to the Great Recession, large-scale biodiesel plant construction surged as Wall Street investors jumped on the bandwagon seeking quick profits. But petroleum prices and global commodity prices soared too, driving the cost of biodiesel feedstocks up to uneconomical levels. When the markets crashed, so did many of those investments. As the price of petroleum plummeted, numerous multimillion-gallon biodiesel plants around the nation were shut down. Others went bankrupt and were auctioned off before they ever opened. The end of the federal tax credit for biodiesel in 2010 was the last straw for many struggling biodiesel ventures (the tax credit was restored in 2011, but expired once again at the end of the year).

The carnage of the Great Recession was not confined to large companies. Quite a few small-scale local biodiesel cooperatives succumbed as well, either to falling oil prices or to the ongoing challenges of working with a large number of volunteers while trying to maintain consistent quality. Others found it increasingly difficult to find free used cooking oil, which for years was the preferred local feedstock for many smaller producers. The most successful smaller biodiesel operations proved to be those that diversified and offered a range of products and services to help them ride out the ups and downs of the market.

The picture for ethanol has been somewhat brighter from a producer's standpoint, but challenging nonetheless. Ethanol has been widely critiqued because of its impact on food production, reliance on federal subsidies, and its generally poor energy return on energy invested (EROEI). But it's here and well established: Over ten times more ethanol (thirteen billion gallons) than biodiesel (one billion gallons) is produced in the United States. In response to ongoing food-versus-fuel concerns, ethanol produced from cellulosic biomass feedstocks has been on the rise (6.6 million gallons at last count), although that's still far below the original EPA target of 500 million gallons by 2012. This shortfall reflects the difficulties cellulosic biofuel technologies have encountered in attracting the capital needed to commercialize, especially during the Great Recession.[1] Nevertheless, the fact remains that a significant number of U.S. corn ethanol producers are local cooperatives that try to keep their dollars circulating in their local communities—and as we've seen elsewhere in this book, a renewable energy source that is ill advised at a large scale may make sense at a local scale. No biofuel is entirely good or bad.

In this chapter, we'll look at a local biodiesel cooperative in North Carolina that is viewed by many as one of the most successful in the nation but nearly succumbed to international market forces; a co-op in Massachusetts that has been impressively persistent and innovative in the construction of a local biodiesel plant; and a locally owned Iowa cooperative that is adding a cellulosic ethanol capability to its existing thirty-million-gallon facility, while trying to increase the value of its co-products and the amount of money it circulates in the local economy.

Piedmont Biofuels, North Carolina

Of all the small, local biodiesel co-ops in the nation, Piedmont Biofuels has been one of the most visible—and enduring. Founded in 2003, Piedmont is a member-owned cooperative of about four hundred members located in Pittsboro, North Carolina. The group has been leading the grassroots sustainability movement in North Carolina by using and encouraging the use of clean, renewable biofuels. Members are entitled to buy biodiesel from the co-op or learn how to make their own. The co-op has six retail outlets for its biodiesel. Piedmont Biofuels also has a biodiesel fuel terminal, the first of its kind in the state, as well as two trucks that it uses to deliver biodiesel in the region, and one truck used to collect waste cooking oil from area restaurants. Most of the biodiesel the co-op sells is for "on road" uses, meaning that the price of the fuel includes all state and federal taxes. A smaller amount of biodiesel is sold for "on farm," home heating, marine, generators, and other "off road" uses without road taxes. Regardless of its end use, all of the biodiesel sold by the co-op meets the ASTM D6751 quality specification.[2]

In addition to selling biofuel, Piedmont consults, designs, and builds small biodiesel reactors, and trains customers on how to use them properly. Local, micro-scale biodiesel production and consumption makes a lot of sense, and the co-op is committed to spreading the knowledge and technology necessary to produce quality biodiesel in small quantities. Piedmont also has a strong educational program. It created the Biofuels Program at Central Carolina Community College, which has expanded to include green building and all types of renewable energy. The co-op hosts workshops on a wide range of biodiesel-related topics, and its employees have taught, demonstrated, and lectured widely around the world. Piedmont founder Rachel Burton is routinely invited to deliver papers on sustainable biofuels at international conferences and events.

In its early years, the co-op made biodiesel at its backyard facility in Moncure. People who wanted to learn how to make biodiesel out of waste vegetable oil became members of the co-op and then joined in with an experienced homebrewing crew. Homebrewing at the co-op is a voluntary activity that has risen and fallen with the enthusiasm of the membership. It is legal to make your own fuel and use it to drive on public highways; in order to stay legal, however, the co-op must pay road taxes on all homemade fuel consumed on behalf of its members. Dealing with regulations has been one of the larger challenges the co-op has faced. "We're out there speaking, blogging, writing books, shooting our mouths off, and as a result we're just way too public to try to do anything under the table," says Lyle Estill, president and cofounder of Piedmont Biofuels. "We've been visited by the IRS three times, and North Carolina Revenue twice."[3]

FIGURE 11-1. Piedmont Biofuels "Industrial" site. Photo by the Abundance Foundation.

In 2006, Piedmont Biofuels made the transition from a small, backyard cooperative biodiesel producer to a small cooperative industrial producer with its Piedmont Biofuels Industrial LLC venture. "After years of success-fully resisting the urge to go into commercial biodiesel production, we finally succumbed," Estill admits. Located in an abandoned alloy plant on the edge of Pittsboro, the multi-feedstock batch process facility is designed to produce about a million gallons of biodiesel annually. However, the new facility has had its share of problems. Opened a year behind schedule and "way over budget," the venture had to be refinanced twice and received a large grant from the North Carolina Energy Office as well. For about a year, the new facility turned out two-thousand-gallon batches of biodiesel steadily for the local market as planned. Then, in 2007, something totally unplanned occurred that put Piedmont's survival in jeopardy.

"That was when the euro was high and the U.S. dollar had gone into the hopper," Estill recalls. "During that period, America was at a discount; that's when the Europeans were shopping in Manhattan and street vendors were accepting euros. So, the Europeans bought up all the poultry fat in the south-eastern U.S., and once they had control of all the feedstocks they went to eleven different biodiesel producers in the area and essentially said, 'Well, we have all the chicken fat now, how would you like to make fuel for us?' Eleven companies, including Piedmont, signed on. Of the eleven, Piedmont is the only one that is still in existence," according to Estill.[4]

"We shipped about 1.3 million gallons of poultry-fat-derived biodiesel to the European Union," he continues. "That lasted for a year. In the summer of 2008, petroleum hit a record high, global commodity markets went through the ceiling, the euro came crashing down, and poultry fat became too expensive to be used. The deal with the Europeans came crashing down as well, and most of the biodiesel partners went down with it. And that was the end of that."

Piedmont Biofuels managed to survive the crash, and returned to making biodiesel for the local market from waste cooking oil. However, the demand for that waste oil had continued to grow, making it more expensive and harder to acquire. And Piedmont's larger industrial facility needed far more of the stuff than it did previously when the co-op was making small batches in its backyard processors. "The plant is running nowhere near capacity," Estill says. "We have a million-gallon facility and we will probably only make about three hundred thousand gallons of fuel this year. We are running at about 30 percent of what it is capable of producing because we are starved for feedstocks."

The main problem is intense competition for the used cooking oil. "Nine years ago, I could jump into any Dumpster in the area," says Estill. "They were overflowing, the lids were open, and the restaurateurs were pleased for us to take as much as we wanted. Now they're watertight, locked, and under contract." The price for fats, oils, and greases remains high due largely to demand in the animal feed market. Piedmont has responded to this new reality with a Partner in Sustainability program for area restaurants. The co-op supplies collection bins (barrels or Dumpsters) to restaurants within a hundred-mile radius of Pittsboro and collects the oil on a regular schedule. The co-op also promotes the participating restaurants to its members and encourages them to eat at these establishments. "We're paying for used cooking oil now, and it's a fierce market," Estill says. "We've probably got a dozen companies vying for used cooking oil in the Research Triangle Park area."

While many larger biodiesel companies were struggling, Piedmont had some noteworthy successes. In August 2008, Piedmont became the smallest commercial U.S. biodiesel producer (and the first in North Carolina) to receive BQ-9000 accreditation.⁵ BQ-9000 is a quality management program administered by the National Biodiesel Board in Jefferson City, Missouri. "We're the smallest BQ-9000 producer on the planet," Estill says. "I think it was a bit of a shocker for the industry when we got that accreditation. They thought that it was only for the big producers like Cargill or ADM."

In addition, Piedmont has an active consulting business. "We have a really well-equipped biodiesel lab, and we do outside testing for people from all over the place," Estill says. "We do field quality consulting and regulatory consulting for other plants in Canada, the United States, and the Caribbean." But Piedmont's biodiesel R&D lab has played a key role in another, increasingly exciting development: enzymatic biodiesel.

Piedmont Biofuels commissioned the first enzymatic biodiesel plant of its kind in the United States in July 2010. This sustainable, scale-neutral biodiesel technology was developed in cooperation with Novozymes (a Danish biotechnology company) to produce high-quality biodiesel using enzymatic catalysis. The new process, which uses enzymes, can make use of low-quality fats and greases that are otherwise hard to transform into biodiesel, while also producing a high-quality, valuable by-product known as technical glycerol.

"When you think of fats, oils, and greases as a spectrum, you start with soybeans, which are easy to work with but expensive," Estill explains. "Animal fats are a little bit cheaper but a little harder to deal with. Then there's used cooking oil, cheaper, but harder still. That's really where the industry ends today. But there are still mountains of fats, oils, and greases in that spectrum, all the way down to trap grease, which is not only free, they'll pay you to take it—but it's extremely difficult to turn it into fuel. That's where enzymatic catalysis comes in; it can handle all that stuff. It's a transformative technology that I think may be a game changer for this industry."

While most other researchers were still working on this problem at the test-tube level, Piedmont already had a thirty-gallon-a-day batch processor set up in their facility. "As a result, we've been taking in fat from all over the planet for testing," Estill says. But enzymatic catalysis has moved beyond the test phase. "There is a twelve-million-gallon-per-year plant being built in Brazil based on our technology, and two of them will be going up in Indiana as well." Consequently, Piedmont's licensing income for the new technology is starting to grow to significant levels.

One of Piedmont's main strengths is that it is a diversified cooperative that sells directly to its members, who collectively own about 450 biodiesel-fueled vehicles. Another unexpected bright spot for Piedmont has been the success of its Eco-Industrial Park where the industrial biodiesel facility (referred to simply as "Industrial") is located. The former alloy plant is now home to a wide range of small businesses, including Eco Organics, a farm co-op; Screech Owl Greenhouse, a sixty-foot hydroponics greenhouse; Piedmont Biofarm's food production space; the Abundance Foundation, which educates the public on sustainability issues; and HOMS, a manufacturer of all-natural insect repellents and pest controls. Other small sustainability-focused start-up businesses keep arriving.

Estill admits that it has been a long and winding road since the co-op was first formed in 2003. "The industry has risen and fallen and risen again and fallen again. Today we get a dollar per gallon of producer credit from Uncle Sam, but that will probably expire in 2012.[6] The last time it expired, in 2010, a lot of biodiesel plants shut down, although we managed to stay open. But we're not making lots of money; we're still a project that's powered mostly by passion. We have been at this for nine years, so at least we're still standing when a lot of others aren't. I guess our greatest accomplishment is a proof of concept of how you might go about fueling yourself. In our view, it's infinitely preferable to have a hundred separate million-gallon plants scattered around the country in small towns rather than a single hundred-million-gallon plant. We're not interested in being the next big fuel monopoly. We're just trying to fuel our community."

Northeast Biodiesel, Massachusetts

If there were a prize for patience and determination, Northeast Biodiesel and its owner, Co-op Power in Greenfield, Massachusetts, would certainly be at the top of the candidate list. This 1.7-million-gallon-per-year community biodiesel plant (which finally appears to be on the verge of opening) has been a long time in coming. The concept for the project came out of a series of meetings beginning in 2002, when a group of organizations—Co-op Plus (a member-owned energy cooperative), the Pioneer Valley Biodiesel Cooperative (a western Massachusetts purchasing group), and the Cooperative

Development Institute of South Deerfield—began to explore their renewable energy options and how they might be strengthened by collaborative effort.

With the help of a grant from the National Renewable Energy Laboratory, the group launched a set of feasibility studies on a wide range of renewable energy technologies, according to Lynn Benander, CEO of Co-op Power and Northeast Biodiesel. One of those studies focused on a regional biodiesel plant. "In our research, our biodiesel plant proposal fit the needs of the biodiesel buying group, and it was something that was easy to understand and could be implemented without a lot of infrastructure upgrades," she says. "It also fit our need to be able to pay living wages to workers, and our need to be able to give at least a low market rate of return to people who let us use their capital. In 2004, we had about 250 people in the room—most of whom had participated in the process for about two years—who voted unanimously to build the biodiesel plant."[7] The new venture, Northeast Biodiesel Company, LLC, was incorporated in 2004. That turned out to be the easy part.

The next phase followed a traditional financing strategy that began with a search to find a venture capital partner. "We found one, negotiated for six months, and at the end we found out that we were going to lose everything that was important to us and not even have a biodiesel plant in our community after their exit strategy was implemented in three years. They told us that we would make a good return on our money," Benander recalls. "When our members looked at just having cash back at the end, they were extremely disappointed because it wasn't at all our primary goal. They didn't want to lose their money, of course, but they wanted biodiesel for the community. Up until then, they felt that as long as they got biodiesel locally that was all that mattered. But they quickly realized that if you didn't own the project, you couldn't guarantee that it would produce an environmentally sustainable product, that the people who worked in the plant would be treated well, and that it wouldn't be sold off for parts after three years." Consequently, Co-op Power stepped away from the $2 million venture capital deal in 2005.

Undaunted, the co-op turned to a traditional bank financing strategy coupled with an innovative private placement memorandum (PPM) with high-net-worth investors who would provide 50 percent of the capital that was needed, and would give Co-op Power a majority ownership in the plant even though the co-op hadn't paid for it. "We weren't following traditional rules, but we were working with people who were committed to long-term community stewardship of this resource," Benander says. "We raised more than a million dollars in financing, and then we went to the banks for their half. But then the banking crisis hit, the biodiesel crisis hit, and we weren't able to get the matching bank financing." As the Great Recession deepened, the co-op continued to seek the bank financing it needed for the project to move forward. Weeks became months. The months dragged on and became years.

Finally, in 2010, the co-op decided to try a new strategy, and announced a "community-based approach" of turning to its members for loans to finance the project. "We went to our high-net-worth investors and asked them if they wanted to leave their money in; and out of the $1.2 million that we had raised, $830,000 of it remained committed, which I think was phenomenal after all those years," Benander says. As of early 2012, $200,000 still needed to be raised in start-up capital from the co-op's 420 members. Nevertheless, the project moved forward incrementally with the funds that were available: The concrete foundation for the plant was poured at Northeast Biodiesel's 26.5-acre site in Greenfield, two large holding tanks for biodiesel storage have also

FIGURE 11-2. Northeast Biodiesel plant under construction in Greenfield, Massachusetts.
Photo by Northeast Biodiesel.

been installed, and the building shell was erected in late February. As soon as the rest of the money is raised, the rest of the plant will be completed. "We have patient capital, and we are not tied to a fixed schedule."

Once the plant is up and running, it will employ fourteen workers who will be turning used cooking oil from more than 125 area restaurants into biodiesel for vehicles and home heating oil blends. Co-op Power has contracted with Holyoke-based ReEnergizer to gather and pre-process used cooking oil from restaurants within an approximately fifty-mile radius of Greenfield.[8]

If Northeast Biodiesel were Co-op Power's only project, it is unlikely that it (or the co-op) would have survived. However, the co-op has had much more success with its many solar and energy-efficiency initiatives. "Energia, a business we built with two community-based nonprofits, has been operating since 2009 in Holyoke doing residential, multifamily, and commercial energy efficiency. It took us six months to set up and is now employing a lot of young people out of Holyoke doing great efficiency work," says Benander. "We've also built our own energy-efficiency division, with $1.2 million in projected sales this coming year. In addition, we've done neighbor-to-neighbor energy-efficiency work, and solar hot water installations, as well as building our vendor network that puts in renewable energy systems for our members. Those are all areas where our story is much more successful, and that's why people have sustained their faith in the biodiesel plant. If it was a stand-alone project, I think we would have all walked away from it."

The lack of easy access to capital has unquestionably been the biggest challenge for the Northeast Biodiesel project. But there have been others. Considerable effort has been required for education and outreach, according to Benander. "We've had to educate the public about recycled vegetable oil and biodiesel, and it has taken a lot of time," she says. "We have also had to spend a lot of effort talking with lawmakers about recycled vegetable oil and biodiesel and how it compares to other biofuels, and why supporting current technology for these fuels is a good idea. Educating consumers and public policy leaders has been very demanding." And, as is the case for many local renewable energy projects elsewhere in the United States, inconsistent public policy has been a real challenge. "The lack of consistent policy has created big shifts in the feasibility of biodiesel projects and made it hard for bankers to assess the risk for investing in a plant like ours. It's been difficult at both the state and federal level."

Despite the many challenges, Co-op Power has accomplished much since it was founded in 2005. The co-op and its growing network of autonomous local organizing councils in Massachusetts, Vermont, and New York are gradually gaining momentum as a regional response to supporting local renewable energy initiatives. "I think building a large network of people who understand the importance of community-based energy efficiency and sustainable energy—and who also understand that who owns it matters—is a huge accomplishment," Benander says. "That we have two hundred volunteers working on projects at any given time, including the development of nine different businesses supporting around a hundred jobs in the region, is another accomplishment. We've also made a big contribution to the local capital scene by creating a very innovative private placement memorandum allowing a community organization to retain long-term ownership in exchange for the short-term liquidity desired by investors. We now have a list of about 150 high-net-worth individuals who are interested in supporting community-based ventures in the Northeast. And then being able to raise hundreds of thousands of dollars from our members has allowed us to make a lot of this happen. All of these infrastructure resources are pretty exciting."

Benander is philosophical about Northeast Biodiesel. "Co-op Power exists to build sustainability and justice, so everything we do is focused on the mission," she says. "No business in our network is too big or too small to fail. As long as there is volunteer spirit and energy, and member interest in getting something done that can make money, we keep gathering resources and we get it done. If at some point the members decided this was not a good idea and didn't want to do it anymore, we would stop. But there is no mission-based reason to stop the development of the biodiesel plant, and we have the resources to keep moving along at whatever speed we can manage. Right now, we are limited by how much money members put in. Whenever members put in enough money, we'll go forward. We're not running by the usual for-profit business rules for development. Normally, you start to lose a huge amount of money by not launching a business, but we don't have a financial burn rate for the biodiesel plant. We have patient capital; and some of our investors are passing up a benefit they might have had from putting their capital somewhere else, but they clearly felt that this could work. They saw what was happening in the industry and what was happening in the banking market, but were willing to wait. There will be a time for all of this, and now it looks like it's time."

Although Benander has been one of Northeast Biodiesel's primary champions, she also praises others who have been part of the team. "I'm very proud to be part of a group of fifty people that have stewarded this project through challenges that would have made all for-profit groups—and most community groups as well—just cut their losses and leave. It's a group that's extremely dedicated and visionary, and I'm honored to work for them."

Quad County Corn Processors, Iowa

Quad County Corn Processors is a locally owned cooperative that runs a thirty-million-gallons-per-year (mgpy) ethanol plant in Galva, Iowa. The cooperative formed in 2000 and raised $8.5 million in private equity to build the first phase of the plant, with most of the shares being sold within a sixty-mile radius of Galva. "Our ownership is all Iowans, and primarily local community folks," says Delayne Johnson, the plant's general manager. The cooperative's membership, currently about four hundred, comprises farmers, local businesspeople, and community supporters.

FIGURE 11-3. Quad County Corn Processors facility in Galva, Iowa. Photo by Quad County Corn Processors.

Construction of the plant was completed in 2001, and ethanol production began in early 2002. Originally built to produce eighteen mgpy, the plant has since been upgraded and fine-tuned and now turns out thirty mgpy. That figure is projected to go up again by roughly two million gallons, thanks to a new patent-pending cellulosic conversion process that Travis Brotherson, Quad County's plant engineer, has been developing for several years. Ironically, the process was an accidental discovery.[9]

"We have a pilot R&D facility," Johnson explains, "and Travis came across a process that converted cellulose into ethanol as he was researching something else. He was able to turn it into a profitable process by making some adjustments to the original development. Quad County is now planning to demonstrate this on a full-scale, continuous basis, and ultimately commercialize it for use in other ethanol plants."

In 2011, Quad County was awarded a $1.45 million grant from the Iowa Power Fund to assist in the construction of its full-scale cellulosic demonstration facility. This was a major step forward for the project, since it had to go through a rigorous technical review prior to approval of the grant. The total project cost is estimated at $6.5 million. The new technology will use the remaining product from the existing ethanol process that would otherwise become distillers' grains (normally a by-product). Although the exact details of the cellulosic process are proprietary, it will be fermentation-based and different from processes being tested by others, according to Johnson. It was the unique nature of Quad County's process that attracted the attention of the Iowa Power Fund, according to Kristin Hanks, program planner for the fund. "This is a smaller project that will be located at an existing ethanol facility and provides another pathway to both get more cellulosic ethanol and provide additional revenue streams," she says. "It was an interesting project for the board from that perspective."[10]

The Quad County strategy also differs from many other cellulosic projects because it focuses on making better use of its existing feedstock, corn, rather than other potential biomass sources such as cornstalks or corncobs, wheat straw, or switchgrass. "We're focusing on corn because the feedstock is already in our plant, and a fair amount of energy has already been expended getting it ready to be converted to cellulosic ethanol," says Johnson. "We also have what we need here to treat the co-product at the very end to become a salable product. So, we can use a considerable amount of our existing plant."

Johnson emphasizes that the new process is not just about producing more ethanol from the same quantity of feedstock. "It's actually creating three value streams from one process," he says. "We're producing cellulosic ethanol, we're producing corn oil, and we're producing a higher value animal feed product which is lower in fiber and higher in protein. Not relying on one value stream also gives us good diversification." Quad County currently processes eleven million bushels of corn annually, and that will not change with the addition of the cellulosic process, according to Johnson. Normally, around a third of each bushel becomes ethanol, a third becomes co-products such as animal feed, and a third becomes carbon dioxide. The new cellulosic process will increase the ethanol yield by about 7 percent.

The planned cellulosic facility has been described as "bolt-on" in the media. "That means that it's not an intrusive process," Johnson explains. "We're able to connect easily to any plant that currently exists. We don't have to remove lots of processing pieces, so there is no downtime; we're able to adapt to any existing facility with just a few minor pipe connections. In our case it will be located in a new building."

In addition to the Iowa Power Fund grant, Quad County has also received a $150,000 grant from the Iowa Department of Economic Development to continue the R&D on the project. Quad County has also applied for a $4.25 million grant from the USDA and U.S. Department of Energy, which if approved will help construct the full-scale demonstration facility. Since the plant is located in an Iowa Enterprise Zone, Quad County may apply for benefits under that program as it gets close to construction. Several other funding sources are being considered, including possible partnerships with other companies. The project is expected to take twelve to eighteen months to complete, and Johnson hopes the full-scale demonstration facility will be operational in 2013. Once the facility is operational, Quad County plans to market the new cellulosic technology to other ethanol producers around the country and around the world.

The development of the new process has had a number of challenges. With assistance from the Iowa Renewable Fuels Association, Quad County is working with the Environmental Protection Agency to recognize corn kernel cellulose as an approved pathway in the Renewable Fuels Standard 2 (RFS2). This would allow ethanol from the new process to be classified as cellulosic ethanol, rather than standard ethanol, allowing the fuel to be used to meet RFS2 usage requirements and to take advantage of federal tax credits. "We believe we'll receive the cellulosic status, but it has been very time consuming," Johnson says. "That's probably been one of our biggest hurdles."

Another hurdle, according to Johnson, is the difficulty of obtaining the capital for a project that is based on a new technology. And the challenge is not having stable government policies. At this writing, incentives for cellulosic ethanol are scheduled to expire at the end of 2012. Lenders need longer periods of time with a stable policy to be comfortable with loaning money. "If we were to build today, by the time the project would be completed the policy would no longer be in existence," says Johnson. "We need consistent policies that last five or ten years, so we understand what the playing field is going to look like."

Despite the challenges, Johnson, who was one of the company's founding board members before he became general manager, is proud of Quad County's contribution to the surrounding communities. "One of the things that we have historically done well is finding ways to add value for our shareholders and our surrounding communities," he says. "This cellulosic process will add value to a feedstock grown in the area and will continue to grow the livestock industry in our area. All of this helps keep dollars circulating close to home, and that's important to us." Quad County sells its carbon dioxide to a facility that converts it into dry ice, which also creates local jobs. "It's a low value stream for us, but we do market it, and it also helps our carbon footprint."

Johnson feels that the local cooperative ownership of Quad County is a good model. "I feel that local control and ownership is a good thing," he says. "It allows us to adapt our operation to benefit the shareholders and the local economy when opportunities arise."

And Quad County's local co-op owners have been supportive of the plant in return. "I think our members have been very supportive over the life of our company. The cellulosic project support is more difficult to measure because of the proprietary nature of the process. We are not able to discuss the details of the process with them, but they are very supportive of any project that helps the company and the industry." Johnson is optimistic about the cellulosic project and the outlook for more like it. "I feel very good about the project," he says. "We're very excited about it."

Geothermal

Geothermal resources could meet a substantial portion
of the nation's energy needs in the 21st century.
In fact, when including geothermal heat pumps (GHPs),
geothermal energy is used in all 50 U.S. states today.
—Bruce D. Green and R. Gerald Nix[1]

As mentioned previously, geothermal energy is divided into two main categories: high-temperature and low-temperature. High-temperature geothermal resources are associated with geologically active regions and are normally used for generating electricity and (at somewhat lower temperatures) for district or other direct heating purposes. However, after a promising start and a century of development in the sector, there are currently only twenty-one geothermal district heating systems (GDHS) operating in the United States, with a total capacity of about a hundred megawatts thermal.[2]

Assessments by the U.S. Geological Survey and others have shown that U.S. geothermal resources can support substantial increases in direct use for heating applications, up to ten thousand megawatts thermal. The main impediments to that growth are primarily social and political (rather than technical), with a complicated legal and regulatory bureaucracy that discourages all but the most persistent developer. On the consumer side, retrofit costs often outweigh the benefits of connecting to a geothermal system; incentives are needed to help reduce these capital costs.[3] Unfortunately, recent budget cuts in federal geothermal program funding have made an already difficult situation even worse.

Low-temperature geothermal resources, by contrast, are available almost anywhere and can be tapped with the aid of a geothermal heat pump (GHP). This strategy is especially useful for heating buildings. Heating and cooling buildings, both residential and commercial, consumes nearly 42 percent of all energy used in the United States today—most of it fossil fuels.[4] Geothermal (often referred to as ground source or geoexchange) heat pumps offer a viable alternative that relies on a local resource. The potential for greater use of geothermal heat pumps for local community projects is enormous.

In this chapter, we'll look at the largest municipal geothermal district heating system in the United States, located in Boise, Idaho; a combined heat and power (CHP) geothermal system at a university campus in Klamath Falls, Oregon, the first of its kind in the nation; and finally, numerous ground-source (geoexchange) systems used in a large collaborative community development located in Oklahoma City, Oklahoma, the nation's largest project of its kind.

Boise, Idaho

At 120 years old, the district heating system along Warm Springs Avenue in Boise, Idaho, is the oldest geothermal system in the nation. When it opened in 1892, the Artesian Hot and Cold Water Company (as it was then called) had two 400-foot-deep wells that had been drilled at the warm springs about 2.5 miles east of town; they produced an artesian flow[5] of 550 gallons of 170-degree-Fahrenheit water per minute (gpm). The wells were connected by a wood-stave pipeline to a new natatorium on Warm Springs Avenue (which remained in business until 1934). The centerpiece of the impressive building was a geothermally heated pool, 65 feet by 125 feet; it also included fifty baths and dressing rooms, a dancing and roller-skating balcony, parlors, billiards rooms, card rooms, a café, and a bar. At about the same time, the wood pipeline to the natatorium was extended farther along the avenue to the downtown. Within a few years, the system consisted of four and a half miles of distribution pipe and served two hundred homes and forty downtown businesses. The wooden pipe was replaced by iron pipe several times, and eventually by asbestos cement piping in 1982. The company's name changed several times as well, and today is known as the Boise Warm Springs Water District (BWSWD). From the beginning, the system has provided only a single line (supply only) distribution system. The cooled water flows into irrigation ditches, storm sewers, and surface drains.[6]

Following the oil shocks of 1973 and 1979, new interest in developing alternative energy sources ultimately led to Boise receiving federal funding for three new geothermal district heating systems, according to Kent Johnson, geothermal program coordinator for the City of Boise. "We already had a known resource, and the federal money helped us to develop it further," he says. "The State of Idaho built one that serves the state capitol building and several of the office buildings surrounding it. The Veterans Administration Hospital on the north edge of town developed a system that serves their campus, and the system that the city owns and operates was the third new one."[7] The city's system is administered by the Boise Public Works Department.

In the early 1970s, the State of Idaho began expanding its Capitol Mall office complex, and rising heating costs prompted a study of the Boise geothermal resource. The study recommended a pilot project. As a result, the state health laboratory was converted to geothermal space heat in 1977. Heating cost savings were immediate, and in 1981 two Capitol Mall geothermal wells were drilled; one was a 3,030-foot-deep production well and the other a 2,150-foot-deep reinjection well. The original water temperature was 162 degrees Fahrenheit, flowing from the production well at around nine hundred gpm. By 1982, nine buildings in the complex were being heated by the geothermal resource.[8]

The Veterans Administration Hospital was established in 1929. The VA began looking at potential geothermal development in the early 1980s and drilled a production well in 1983 that was capable of producing up to 1,245 gpm of 161-degree water. An injection well was drilled in 1987, and the system came online in 1988. All thirty of the VA's buildings were connected either directly or indirectly to its geothermal district system, which provides both space heat and domestic hot water. The facility was originally served by a steam plant, and all of the buildings have been converted to geothermal hot water; a hot water boiler provides extra heat to meet peak demand and offer some backup for the geothermal system. The well pumps are connected to a backup power supply. Altogether, the geothermal system supplies heat for 436,000 square feet of building space. The original payback time for the system was estimated at eight years, but the actual payback was achieved in five years. In 2002, the system won an Energy Star award from the Environmental Protection Agency.[9]

Unlike the old BWSWD system, all three of the new district heating systems were constructed with a supply line *and* a collection line—but only two of the three new systems had injection wells to return the water to the

FIGURE 12-1. Geothermal supply and collection water lines being installed in the city's district heating system. Photo by Boise Public Works Department.

aquifer. "The state system had an injection well from day one, the VA system had an injection well when they came online, but the city system did not have one," Johnson says. The cooled water in the city system was disposed of in the Boise River. Before too long, aquifer levels in the underground geothermal resource began to drop significantly.

"There was a lot of discussion about what was happening: studies, lawsuits, and a moratorium in 1988 by the Idaho Department of Water Resources that limited further expansion of the systems," Johnson continues. "The studies indicated that an injection well for the city that would return the water to the geothermal resource might solve the issue, so that generated another round of studies and modeling to determine a good location for an injection well." Finally, in the early 1990s, the city was able to get federal funding for the injection well. After some additional studies, it was drilled, connected to the collection line, and finally put into service in early 1999. The city now injects 100 percent of the geothermal water back into the aquifer. "The recovery was obvious from the beginning, but it wasn't until 2010 that the geothermal resource had recovered to where it was back before the three new systems came online in the 1980s. It's been good for the resource, but I think it's also been good for us to learn how to work together to start to think about what the long-term use of the resource will be, so we can use it to its capacity without hurting it or each other." The four systems withdraw approximately 775 million gallons of geothermal water per year to heat more than two hundred homes and around eighty-five government buildings.

With the recovery of the geothermal resource, the users have come to an agreement (under the provisions of the moratorium) to allow the city's production ceiling to be increased; the city is now expanding its lines to the Boise State University campus. The university's Morrison Center, Multi-Purpose Classroom Building, Interactive Learning Center, Math and Geosciences Building, and Center for Business and Economics Building are all part of the first phase, which is scheduled for completion in early 2012. The Administration Building, Student Union Building, and Center for Environmental Sciences and Economics are among the buildings that will be connected during phase two beginning in the summer of 2012. When completed, approximately 625,000 square feet of building space at the university will be heated from geothermal energy. A majority of the funding for the project is coming from federal appropriations. Excluding the university, the present city system serves sixty-four institutional or commercial customers, heating approximately 3.9 million square feet of building space (about half of the customers are privately owned buildings).[10]

With its high density of buildings in a fairly compact area, the university will be a welcome addition to the city's geothermal system, which features relatively long distribution lines and relatively low customer density in some parts of the system. "It's really popular to be using geothermal heat on a college campus these days; even more than in the general community," Johnson says. "We surveyed our existing customers to find out how they were feeling about the service, and I was surprised because there wasn't a single building owner that said anything about it being renewable or green—it was all about cost. Even potential customers who are excited about geothermal heat because it's renewable want an inexpensive service."

Distribution line maintenance for virtually all geothermal district heating systems is a headache. "Hot water under pressure causes problems with metal piping because the warm pipe corrodes very fast in the ground," Johnson explains. "So you either have to isolate it from the soil or use a different material. Originally, the city system was built with asbestos-cement pipe, which is a fairly good application for that type of pipeline because it can stand the heat and the pressure. However, the metal fittings used to connect the older pipe create an ongoing maintenance problem. With that type of pipe no longer available, our next choice is a fiberglass-reinforced plastic pipe, which is a pretty expensive industrial type of pipe material that costs between $50,000 and $100,000 a block—and that's just for the pipe in the ground. We pay a lot for that pipe, but we justify it because we don't have to worry about corrosion. However, it doesn't pay off unless we have a lot of customers on the line, so that's a real challenge."

Geothermal district heating is admittedly not for every community. Even assuming that a location has the necessary geothermal resource available, the high initial capital costs can be daunting. "Our initial backbone was built with federal funds. We probably would not have gotten the system started in the first place without those federal grants and supports," says Johnson. The decline in federal funding for geothermal district heating systems in recent years has not helped this situation. The funding that was included for geothermal in the Energy Independence and Security Act of 2007 was mostly focused on electricity production and did not support district heating projects.

Johnson is cautious about the prospects for future expansion of the city's geothermal district heating system. "I think it will continue to grow slowly, but I don't think there will be anything dramatic," he says. "I think this expansion to Boise State is the largest we'll see. One of the reasons is that we have many energy options here, especially inexpensive electricity and natural gas; so the first thing people want to know is how much it costs. Unless something else happens in the energy market, I don't see a major shift."

Oregon Institute of Technology, Klamath Falls, Oregon

The Oregon Institute of Technology (OIT), located in Klamath Falls, Oregon, has an impressive claim to local energy fame. Its recently installed geothermal combined heat and power plant is the first in the nation, and the first in the world at a university campus to heat its buildings *and* generate electricity from the same geothermal resource directly below. OIT's main campus is located in the high desert country just east of the Cascade Range in the southern part of the state, twenty-five miles from the California border. The campus has been heated with geothermal water since 1962 when OIT was moved from a previous location, primarily to take advantage of the significant geothermal resource that supplies space heat to around a thousand residences and other buildings in Klamath Falls.[11]

The system began with three geothermal wells, one around twelve hundred feet deep and two others around eighteen hundred feet deep. In the original design, the 192-degree-Fahrenheit geothermal water was used directly in all of the building's mechanical systems. This simple design caused a number of problems, however, because the geothermal water contained a small amount of hydrogen sulfide, which damaged control valves, solder, and copper piping. The water now flows into a settling tank that removes fine-grained sediment, and then flows by gravity to each building on campus where the heat is transferred to cold secondary water through a stainless-steel heat exchanger. The heat exchanger isolates the geothermal water from the building heating systems and protects them from corrosion. The geothermal water is injected back into the ground through the two eighteen-hundred-foot wells. The geothermal water provides all of the space heat and domestic hot water for the sixteen buildings on campus totaling approximately 818,200 square feet. The campus also has 40,400 square feet of geothermally heated sidewalks, stairs, and handicap ramps (for melting winter snows). OIT boasts around $1 million every year in energy savings.[12]

In 2003, the first proposal to use some of the heat energy from the existing geothermal water to operate a small, moderate-temperature power plant was made to OIT Facilities Services. The combined heat and power idea was well received, and an engineering firm was hired in 2005 to evaluate energy use on campus. The firm concluded that there was enough heat energy from the existing geothermal water to run a two- to three-hundred-kilowatt power plant and still heat the campus. A team was set up to write a proposal and solicit bids for the power plant as well as the auxiliary equipment needed for the project.

In 2006, the only company that offered a moderate-temperature geothermal power plant in the output range needed was United Technology Corporation

FIGURE 12-2. Oregon Institute of Technology's 280-kilowatt geothermal power plant. Photo by OIT.

(UTC) of East Hartford, Connecticut. UTC, in collaboration with the U.S. Department of Energy, had recently developed a two-hundred-kilowatt geothermal power plant (the first of its kind in the nation) for Chena Hot Springs, a small resort community in rural Alaska. The operational performance reports from the Alaska installation were positive, and in January 2009, OIT decided to order a slightly larger 280-kilowatt unit for their campus. The unit arrived two months later and was installed in OIT's existing heat exchange building; the new power plant was dedicated in early 2010. "Because we are a university campus, the primary goal was to use the power plant as a demonstration site," says Tonya "Toni" Boyd, senior engineer of the Geo-Heat Center at OIT. "But because this is not something we do every day, it's been a steep learning curve for us."[13]

In addition to normal building permits from the county, a number of other approvals and permits were needed for the project to move forward. The Oregon Department of Energy required a state energy-efficiency design certification. Water rights certificates and Underground Injection Control (UIC) permits had to be reviewed and revised to include power generation as a use (the City of Klamath Falls passed an ordinance in the early 1990s that requires pumped geothermal water be reinjected into the ground). In order to generate electricity while interconnecting with the power grid, approvals were required from OIT's local utility, Pacific Power. OIT obtained approval, although the price per kilowatt-hour that the institute receives for its power being fed into the grid is less than what it pays when it draws power from the grid.

A number of funding sources were tapped for the project. The Energy Trust of Oregon assisted OIT in hiring consultants to help guide the project team through the power plant development process. The consultants were particularly helpful with the interconnection agreement and power purchase agreement, according to Boyd. The Energy Trust also provided a $487,000 incentive, and Pacific Power awarded a $100,000 Blue Sky Grant. In addition, OIT received an Oregon Business Energy Tax Credit of $254,148 after the project was completed. Some additional money came from OIT's own capital projects funding. The total cost of the power plant and associated equipment was $1.1 million; it now supplies about 10 percent of the electricity needed on campus.

The power plant is designed to use about 15 degrees Fahrenheit of the available heat off the top of the geothermal water for electricity generation, leaving water of about 177 degrees ($192 - 15 = 177$) to heat the campus. The power plant needs a minimum of six hundred gallons per minute to operate. During warmer weather, when there is less space heating demand, more geothermal heat can be used by the power plant to generate additional electricity. Very few problems have been encountered since the power plant came online. Nevertheless, the OIT facilities staff continue to adjust the system to make it as efficient as possible while balancing the needs for both heat and electricity.

But that's not all. In 2009, in an effort to find hotter geothermal water, OIT drilled a much deeper (fifty-three-hundred-foot) well on campus. The water turned out to be only slightly hotter (196 degrees Fahrenheit) than the existing wells; however, the new well produced *a lot* of hot water. "One good thing about the deeper well is we have a lot of flow," Boyd says. "We did a water rights test and got twenty-five hundred gallons per minute that we can use." Based on the test, OIT sent out a request for proposals and eventually awarded a contract for a custom-designed, two-megawatt power plant to Johnson Controls. A twenty-five-hundred-foot injection well for the larger project has been drilled, and the permitting process continues. "We still have to deal with Department of Environmental Quality regulations for the injection well for an Underground Injection Control permit. We're also working on the interconnection and power purchase agreement. In addition, we've installed a pipeline from our deep well up to where we are going to put the larger power plant, and we're hoping that by the end of 2012 we'll have it online." The total cost of the deep well and the two-megawatt power plant will be around $12 million. The combined output from the two power plants should cover about 75 percent of the total electricity needed on campus.

Although the two-megawatt power plant project has been designed primarily for electricity generation, there is CHP potential as well, according to Boyd. "After we use the water in the power plant and send it via pipeline to the injection well it's still pretty hot, so it could be used for other things as well," she says. "We are hoping that if greenhouses or other agricultural operations are developed on campus, they're sited near the pipeline so they can make use of that excess heat."

Since OIT offers numerous engineering courses, the new geothermal power plant provides students with excellent examples of real-world application of their academic work. "We try to keep the students informed about the project because we have a renewable engineering program here," Boyd says. "They can go up to the power plant while we are working on the project and get a good idea about what's going on." The projects offer OIT other educational opportunities. "Besides being able to generate the power and have the demonstration site, we can also help other people develop small projects of their own," Boyd adds. "We've gone through the process, and we can help them get over some of the hurdles they might encounter when they do this."

The potential for more projects is considerable. There are over four hundred communities in sixteen western states located within five miles of a geothermal resource, according to Boyd. "Not all of them have the potential for power generation, but at least fifty-eight do. And several other communities in Oregon are currently working on or looking at power generation projects."

Boyd offers some advice to other communities that might be considering a geothermal project of their own: "You really need people who are dedicated to the project to get it going. You can't just decide to do a power plant and have it happen right away." She also suggests hiring knowledgeable consultants to help with the permitting and approval process. "Most people are not familiar with interconnection and power purchase agreements and the other legal technicalities, so it's nice to have someone who knows what they are doing to help you out with them. At the same time, that person might not have the geology background that you need, so it's important to get the right person for the different kinds of expertise that you need."

Despite all the time and energy required to ensure a successful project, Boyd is happy with the results so far. "I feel really good about it," she says. "And it's also been a lot of fun."

Hope Crossing, Oklahoma City, Oklahoma

Hope Crossing, located in northeast Oklahoma City, may look like other suburban developments with its compact brick homes, tidy yards, and clean streets. But that's where the similarity ends. This is because Hope Crossing is one of the most energy-efficient developments in Oklahoma, or anywhere else in the nation for that matter: It offers comfortable, easy-to-maintain homes that feature ground-source geothermal heat pumps for heating and cooling, as well as many other energy-saving features. Even more remarkable is the fact that these quality homes are owned by low-income families that never dreamed they would be able to have a home of their own.

Hope Crossing is a nonprofit enterprise developed by Central Oklahoma Habitat for Humanity on fifty-nine acres donated by a local real estate investor, Stephen Hurst, and his partners in TexOk Properties, LP. The land, valued at about $550,000, was an open field grazed by a few cattle, according to Ann Felton, chairman and chief executive of Central Oklahoma Habitat for Humanity. "We had known for a number of years that we were going to have to become developers because of our aggressive building schedule, but quite frankly we just didn't have the money to buy the land plus the millions of dollars needed to develop it. So, this was a great blessing to receive the land in a donation."[14] Once the land was secured, the necessary infrastructure, including roads, water, sewer, drainage, and electricity, was gradually installed over a number of years with the support of a large number of community donors and project partners. "None of this would have been possible without our partnerships," Felton adds.

Launched in early 2006, the multiphase, 217-home project located at NE 83rd and Kelley Avenue features three-bedroom, two-bath homes averaging

1,250 square feet. The first phase, completed in late 2006, included the construction of thirty-one homes from a selection of five different designs by Oklahoma City architects Fillmore Design Group. The development now has more than 170 energy-efficient houses that are appraised at about $125,000 but only cost around $85,000 to build.[15] This bit of financial magic is the result of around three hundred hours of sweat equity from the prospective homeowners, as well as the efforts of an estimated 22,473 community volunteers who have helped with the construction over the years.

Qualifying families purchase their homes from Habitat for what it costs to build them. No profit is added, and financing is with zero-interest mortgages featuring no down payment and affordable monthly payments based on income. The homes' high energy efficiency offers many environmental benefits and slashes utility costs up to 50 percent, another major benefit for the low-income homeowners. Once all the homes have been completed, probably in 2012, Hope Crossing will be the largest green-built Habitat community in the United States. It will also have a 2.5-acre park with playground equipment, curving paths, and hiking trails for residents thanks to the generosity of a number of donors.

The use of ground-source heat pumps at Hope Crossing was the result of a partnership with ClimateMaster of Oklahoma City, a manufacturer of high-efficiency, ground-source heat pumps. Habitat had established the partnership on a previous project. "We did ten houses in our Spencer Subdivision with ClimateMaster, and we were very satisfied with the performance of the heat pumps," says Aaron McRee, Hope Crossing construction manager for Central Oklahoma Habitat for Humanity. "ClimateMaster has been very good to us over the years; they donate the units to us and we pay for the loop drilling and the installation of the units in the houses. It's probably one of the best things that we could have done for our homeowners."[16]

Early on in the project, Habitat decided to continue its successful use of heat pumps for all of the homes in its Hope Crossing development. By installing high-efficiency ground-source geothermal heating and cooling systems, Habitat offers homeowners the benefits of more comfortable homes, and more affordable utility bills as well. "The geothermal units that we use are rated at twenty-seven EER [energy efficiency ratio].[17] I don't think there is anything out there that is more efficient than that," McRee says. "The government minimum is a thirteen SEER [seasonal energy efficiency ratio] on a typical conventional system, so the geothermal systems we install are about twice as efficient as what a conventional system would be."

Habitat incorporated a number of other time- and money-saving strategies with the geothermal installations to help keep costs as low as possible. The heat pump itself has all of its components combined in a single unit that is installed in a utility closet in the home, eliminating the need for a second unit located outside. "The outdoor compressor is often a target for thieves or vandals during construction, so that's an advantage of the systems we use," says McRee. In addition, a single 400-foot geothermal heat exchange loop is located directly under the floor slab in the garage instead of the typical practice of drilling two 200-foot loops 20 feet apart outside of the home in the yard. The latter approach would require an additional excavation step to connect the heat exchangers to the piping in the house.

But the carefully engineered, integrated design of the homes in Hope Crossing results in even greater savings with the geothermal systems, according to McRee. "It's important to note that because of the insulation and other things we do to make the homes tight, we were able to downsize the geothermal system from what it otherwise would have been in a conventional home. Typically, a home this size would have a 2.5-ton system, but we were able to downsize it to a 2-ton system.[18] Of course, that also benefits the homeowner because the smaller the size, the less energy it uses."

The homes' careful engineering offers additional benefits, especially considering the many different partners and volunteers that have been involved in the various phases of the development's construction over the years. "On a typical house, the contractors are the ones who decide what size diffuser and return air ducts will be used and where they will be located," McRee explains. "But we have an engineer who has done all of those calculations and drawings for us, so the installation will be the same no matter who the contractor is. The amount of air going to each room is already calculated, and all of the joints on our air ducts are carefully sealed so there is very little loss from the geothermal unit to the entry point for the room."

All this engineering does not mean that the techniques and materials used at Hope Crossing haven't evolved and improved over time. They have. "We've learned a lot from the start of the project up to now. We keep improving what we are doing," says McRee. "We started out with fiberglass insulation and then switched to foam. We've upgraded our windows, and many other small things that add up to a lot. We've tried to build the most energy-efficient and maintenance-free house as we can with the resources that we have to work with."

In a collaboration between ClimateMaster and Oklahoma Gas and Electric (OG&E), two of the houses in the development have solar PV panels installed on their roofs. "We selected two of the earlier houses and installed twelve panels on each of them," McRee says. "This was a test to see how well they would perform. It's not enough to make them zero-energy homes, but it should be enough to pay their heat and air bills." The solar panels have reduced grid energy consumption for the two homes between 60 and 80 percent.

OG&E has also collaborated with the International Ground Source Heat Pump Association to conduct a two-year test on various ground loop designs, according to McRee. "We identified ten of the older geothermal houses that we had built, and they disconnected the original geothermal ground loops and installed different loop designs with different depths and different piping to see how they would compare. The study is designed to see which ones perform the best, with the overall goal to find the least-expensive and best method to use." The test began in the summer of 2011, and the results are anticipated in 2013.

A project this size that spans so many years is not without its challenges. The crash of the housing market and the tightening budgets of the Great Recession did not help. "But what really hurt was when the state tax credits were eliminated," McRee says. "About sixty homes in the earlier phases of the project were able to take advantage of the $4,000-per-house state tax credit before a moratorium was imposed in June 2010 due to budget cuts. The tightening of budgets also suspended the LEED certification for the homes in Hope Crossing, according to McRee. "Up until 2010, all of our houses were built to LEED standards and were LEED-certified. The cost to get the actual certification was covered by grant money, which dried up, so we opted not to get the houses certified, even though we were building them to the same standards. We couldn't see spending several hundred dollars more to get a piece of paper that said they were certified when we know how well they are built."

Despite the challenges, the project has retained strong support from the city and from Habitat's many partners and volunteers. "Donations have been a bit lower due to the economy being down, but we've got a lot of support from our city council and all the city departments," McRee says. "And our volunteer numbers have never been higher; it's been extraordinary." Ann Felton agrees: "We've had an incredible amount of support."

Of course, the most important feedback on Hope Crossing is from the homeowners. "The ones that I've spoken with are extremely happy with their homes," says McRee. "Many of them came from smaller rental properties that had electric bills three times higher than what they are now paying in Hope Crossing. I hear it often; they stop by and tell me how happy they are even two years after they've moved in."

McRee is enthusiastic about the project. "I've been here since the beginning, and to see it from where it started as an empty field to where it is now is really amazing," he says. "To know that this is the largest Habitat subdivision of its kind in the nation, and that it's probably the most energy-efficient subdivision in the nation, is very exciting. It's a great feeling to know that we've all been a part of this."

"I feel very satisfied with the project," Ann Felton adds. "I've been involved with Habitat for twenty-one years, and I've seen this grow to the point where we are able to offer so much more for our families. I've also been able to watch the families grow and prosper. I've seen some of the children graduate from high school and go on to college, or the military, or some other job. It's been very satisfying, and a great blessing in my life as well."

Based on the success of Hope Crossing, Habitat is already planning its next large project. "We are looking for property that we can build on," McRee says. "And we have no intentions of moving away from geothermal. It's been good to us and better for our homeowners, so we intend to stay with it."

Exceptional Community Energy Initiatives

Never doubt that a small group of thoughtful, committed citizens
can change the world: indeed it's the only thing that ever has.
—Margaret Mead

We've looked at many inspiring examples of what people have accomplished in communities across the nation. But there are, of course, a great many others in the exciting and rapidly changing world of community energy, including ones that take a broader approach to building their community's energy resilience. In this chapter, we'll look at three communities that have initiated a wide range of local energy and efficiency initiatives that extend well beyond a single renewable energy technology or efficiency program.

One of the biggest challenges for almost any community is coordinating many different local (and often regional) initiatives into a coherent and complementary program that achieves its goals. It's an especially important issue in a time of declining budgets and limited resources. The City of Gainesville, Florida heads our list with a series of initiatives that have unquestionably made it a national leader in renewable energy and energy conservation. Next is the City of Newburyport, Massachusetts—aided by a diverse group of local organizations—which has instituted a broad series of initiatives intended to make it a net-zero-energy city by 2028. Last, but by no means least, we'll look at a small town in Bavaria, Germany that has pulled together a series of community-based renewable energy and sustainability initiatives, with spectacular results.

City of Gainesville, Florida

By almost any measure, Gainesville, Florida, has become a national leader in renewable energy and energy conservation. A relatively small community (population 125,000) that's also home to the University of Florida and Santa Fe College, Gainesville has been working on renewable energy initiatives for many years. By the end of 2011, the municipal utility that serves Gainesville had 9.8 megawatts of solar photovoltaic systems installed. That statistic puts Gainesville ahead of California in terms of kW per person of installed solar PV—all the more remarkable since California has had a number of programs for solar PV in place since the early 2000s.[1]

This extraordinary accomplishment is largely the result of some very progressive energy thinking on the part of the city and the Gainesville Regional Utilities (GRU) that serves area residents. GRU was the first municipal utility in the nation to implement a European-style feed-in tariff (FIT) for solar PV after previous efforts relying on grants, tax credits, rebates, and net metering failed to make much headway.

"They are unquestionably the leader in the state and the country, and they deserve every penny of economic reward that has flooded into the city as a result," says Mike Antheil, executive director of the Florida Alliance for Renewable Energy (FARE) in West Palm Beach. "I think the feed-in tariff has been a tremendous accomplishment, and it has put the city on the map within the renewable energy industry. When I talk to people across the country about financing renewable energy projects, they aren't interested in Florida in general, but they all want to know what is happening in Gainesville, and they would trip over themselves to finance one of the contracts. Millions of dollars have poured into Gainesville from all over the world, and it's been a huge economic boost."[2]

While Antheil gives former Gainesville mayor Pegeen Hanrahan (2004–10) and the city commission at the time much of the credit for implementing the FIT, he says that Ed Regan, GRU's assistant general manager for strategic planning, was the key driving force. In late 2008, Regan visited Germany to see firsthand how that nation's feed-in tariff worked. "Ed came back from Germany with an understanding of the policy and how it could work here in Florida," Antheil says.

Regan first came to GRU in 1979, when he was hired to set up the utility's conservation programs. Over the years he has been involved in a wide range of energy-efficiency and conservation initiatives. Regan says that on the trip to Germany he encountered a whole new range of possibilities that made sense to him. His report back to city officials apparently made sense to them as well. "When I came back with the German model, I was surprised that the city commission really liked it, and so we adopted it," he says. The vote by the utility's board of directors, the Gainesville City Commission, was unanimous.

The effects of that decision have been dramatic. "It was huge in making our solar program as successful as it has been," Regan continues. "In 2008, when I made the FIT presentation to the city commission, we had maybe 350 kilowatts of solar after a couple of years of solar rebates and net metering. Here we are, three years later, with 9.8 megawatts of solar installed and another 3.7 megawatts under construction, which is huge for a community our size." (In late 2008, the entire state of Florida had only two megawatts of solar PV installed.) The FIT was more attractive to solar investors than were traditional solar rebate programs because it guaranteed that GRU would buy all of the electricity produced by the PV systems at a fixed rate for twenty years, according to Regan. "It offers investors a reliable and predictable source of income."

But that's not the only initiative that sets Gainesville apart. In 2003, GRU had been looking at electricity generation options and floated the idea of

FIGURE 13-1. Some of the many solar PV arrays on commercial buildings in Gainesville, Florida, that were encouraged by the implementation of the city's first-in-the-nation feed-in tariff.
Photo by Gainesville Regional Utilities.

a new coal-fired power plant to provide baseload power. "That started a huge discussion in the community about whether we were doing enough with conservation and renewable energy," Regan recalls. "We'd been doing renewable energy in a small way—we gave rebates mostly for solar water heating—and we had the first green energy pricing program in Florida; but the community wanted to know if we could do more." This led to a long series of public discussions, along with independent reviews, that eventually resulted in a proposal for a one-hundred-megawatt biomass plant that would provide GRU with long-term baseload power generation.

"We spent a long time thinking about and discussing biomass," Regan says. "We are one of the major paper pulp-and-timber-producing regions in the country. If you drive around our area, you see these big piles of slash left over from the harvest. We did some preliminary research that indicated this resource was huge, so that's how we got into the biomass picture."

The Gainesville Renewable Energy Center (GREC), as the waste-wood-fired, electricity generating project is called, is located about eight miles northwest of downtown Gainesville. GRU has entered into a power purchase agreement with American Renewables, the project developer, for all of the power produced by GREC for thirty years. The $500 million center will consist of a wood fuel handling system, a boiler featuring advanced combustion and emissions control technology, a steam turbine generator, and auxiliary support equipment. The project will require about one million tons of fuel annually sourced within a seventy-five-mile radius of the facility. This area encompasses over 5.5 million acres of timberland, more than enough to fuel the project according to an independent forestry consultant. GREC will comply with Florida Division of Forestry Best Management Practices, and has agreed to additional standards to foster overall forest health. Construction is already under way; the project is expected to be operational by 2014.[3]

While there has been some opposition to GREC, most local residents favor the biomass project, according to Regan. In addition, GREC has received support from the Southern Alliance for Clean Energy (SACE) after a thorough review of the project details. "The agreement with the biomass plant is for a fixed price for thirty years, and most people understand the value of that," Regan says. "In addition, we won't be sending a lot of money somewhere else since the plant will create a lot of jobs in the local area, and people are excited about that too."

But that's not all. GRU (which also provides water, wastewater, natural gas, and telecommunications services) has had an active landfill gas program as well. For a number of years, GRU partnered with Alachua County's Southwest Landfill, where it generated baseload power until 2007 when the usable gas resource was depleted. In January 2009, GRU began to purchase electricity from G2 Energy, LLC, which owns a landfill gas power plant at the Marion County Baseline Landfill. The three-megawatt plant generates about two thousand megawatt-hours of electricity per month, enough to power more than twenty-one hundred homes for a year. "What's good about the Marion County landfill is that it's not closed yet, so as cells deplete, they will be adding new cells.[4] We're very satisfied with that," says Regan. "We really learned our lesson on our previous project, because by then we realized that there was no way we could do this as effectively as the private sector: Even though we have tax-exempt financing capability, we can't take advantage of any of the federal incentives."

GRU and the city have a wide range of other environmental and conservation initiatives, according to Regan. "Our conservation programs are huge," Regan notes. "In 2005, the city commission signed the U.S. Conference of Mayors Climate Protection Agreement and tasked us with meeting the Kyoto Protocol targets for carbon reduction. The city optimized all of the traffic signals, and it's estimated that the consumption of gasoline here has gone down 15 percent as a result—a huge carbon reduction." GRU has found further carbon energy savings and carbon emissions reductions by using the biosolids from its wastewater operation as a fertilizer, and through water conservation programs such as a rebate for the installation of moisture sensors to control irrigation systems. And GRU still offers rebates for solar water heating as well.

Regan is very pleased with Gainesville's many successful renewable energy and conservation initiatives. "I'm very proud of our accomplishments," he says. "I think our community has discussed, addressed, and dealt with issues that have been debated at the federal level for who knows how long; we just decided what we were going to do, and we got it done."

Mike Antheil agrees. "Ed Regan knew that they could do it in Gainesville; he brought the FIT proposal to the city commission and the mayor, who picked it up and really ran with it. They pushed hard for it because they understood what many others in the state still don't understand: If a small local group gets together and sets their minds to do something, they can actually get it accomplished."

City of Newburyport, Massachusetts

Described as a "seaport for all seasons," the City of Newburyport, Massachu-setts (population around 17,500), is located at the mouth of the Merrimack River about thirty-five miles northeast of Boston. Famed for its historic downtown, museums, arts events, parks, beaches, and bird-watching, the city is also gaining a reputation for its environmental and sustainability initiatives. In December 2010, Newburyport was designated a Green Community by the Massachusetts Green Communities Program and received a grant of $155,000 to fund conservation measures at the city police station and city hall. Newbury-port joined eighty-five other towns and cities in the state in sharing around $19 million in grants in recent years aimed at reducing energy use in municipal and school buildings, establishing power purchase agreements that enable renew-able energy generation, adopting the latest building codes, and much more.

But even before its Green Community designation, Newburyport, along with other local groups and businesses, had been involved in projects and initiatives aimed at making the city and surrounding area more sustainable. In 2008, then-mayor John Moak formed an Energy Advisory Committee (EAC) made up of people who work in the renewable energy sector to help the city reduce its energy consumption and shift to greater reliance on renewables. "Its role is to advise the mayor on anything energy-related, which takes in quite a broad scope of initiatives: from solar panels to city lights to cars and other things with links to energy," says Niall Robinson, a member of the committee. That scope even extended to wave energy. The EAC collaborated with the Newburyport Waterfront Trust and Resolute Marine Energy Inc. of Boston to test a wave-powered energy converter in January 2009. After several attempts, the test had to be postponed due to extreme weather and technical problems, but it eventually provided some useful information that encouraged Resolute to continue its research and development work (more on that later).

In September 2009, with guidance from the EAC, the city had a five-hundred-kilowatt solar array installed on the roof of the Rupert Nock Middle School, one of the largest municipal solar PV projects in the state at the time. In addition, the city has supported the installation of more bike racks, greener vehicles, energy audits of government buildings, mixed stream recycling,

FIGURE 13-2. An aerial view of the solar PV arrays on the roof of the Nock Middle School in Newburyport, Massachusetts. Photo by Paul Sieswerda.

membership in the ICLEI–Local Governments for Sustainability, and much more. Newburyport is also working on a ten-year revision of its city plan, and a wide range of sustainability initiatives are being incorporated into the plan.

"We have a long-term goal to be energy-independent by 2028 by becoming a net-zero energy city," Robinson says. "We hope to achieve that goal through projects like promoting weatherization and solar panels, as well as through buying 60 percent of the city's electricity from a solar array that will be constructed near here." The proposed six-megawatt True North, LLC solar farm project will be located on forty-three acres in the adjacent Town of Salisbury. The project will divide its electricity output among Newburyport (44 percent), the Triton Regional School District (45 percent), and Salisbury (11 percent), and Newburyport will receive a discount of 10 percent on its share of the electricity for twenty years.[5]

Newburyport also offers its residents a Carbon Challenge. This energy awareness initiative involves a short online survey of energy use coupled with a free energy audit through MassSave (an initiative sponsored by gas and electric utilities and energy-efficiency service providers in Massachusetts). The audit package includes free compact fluorescent bulbs as well as thousands of dollars of rebates and interest-free loans to help with the home weatherization project developed from the survey. "The Carbon Challenge does two important things," Newburyport mayor Donna Holaday said at the launch of the initiative in April 2010. "It quickly shows you how much energy you consume and a number of ways to reduce your consumption. Then, your energy savings are translated into dollar savings—and those savings are no small thing."[6]

But the sustainability initiatives extend well beyond the municipal government. In a separate initiative, a local woodworking business located in an industrial park installed a six-hundred-kilowatt wind turbine in January 2009. The 292-foot-tall turbine was praised by city and state officials at the ribbon-cutting ceremony, but subsequently drew complaints about noise and shadow flicker from neighboring residents. The project had strong support from the EAC. "There have been times when we have not communicated with the public as effectively as we might have," Robinson says. "We got a lot of pushback on the wind turbine project after it was up and running. One lesson we learned was that we should have hired somebody to present the pros and cons of the project up front."

Another local organization, the Greater Newburyport Eco Collaborative, has been active in promoting sustainable practices, environmental initiatives, and educational events. Formed several years ago, the group is a collaboration among the Greater Newburyport Chamber of Commerce, the City of Newburyport, and numerous environmental organizations. The Eco Collaborative funded the Carbon Challenge and supports the Newburyport CleanTech Center (NCTC), which encourages the advancement of start-ups (as well as established companies) that offer marketable solutions to environmental challenges. The center also provides resources to assist with start-up issues such as strategic planning, financial preparation, legal and regulatory issues, sales and marketing, and team development.[7] In April 2010, the NCTC signed an agreement with Resolute Marine Energy, Inc. to support the development of Resolute's operational prototype for a wave energy converter.[8]

While all of this activity is exciting, it is not always well coordinated. "One of our challenges is that there are quite a few groups within our community that are working toward similar goals," Robinson says. "What we need is an umbrella organization or some other way to bring these different groups together so we are all playing the same tune." In addition to his work on the Energy Advisory Committee, Robinson was also a founding member of Transition Newburyport, so he is able to bring that group's perspective to the committee. To an extent, Transition Newburyport has begun to act as a facilitator for some of these community efforts.

Transition Newburyport—part of the international Transition movement for building local resilience (see http://transitionnetwork.org)—was officially organized in March 2009 as the first Transition initiative in the state, and the twenty-first in the nation. Among a wide range of activities, the group has organized "resilience circle" study groups that meet to study Transition concerns such as climate change, peak oil, and economic contraction. "There is a lot happening, and we are trying to partner with others who are already doing good work," says Conrad Willeman, a Transition Newburyport organizer. "We try to identify where the gaps might be to see if we can catalyze some activity in that area. We are working with existing organizations, church groups, community gardens, the farmers' market, the Greater Newburyport Eco Collaborative, resilience circles, and others who see us as encouraging a new spirit of community that has been missing up to now."

Niall Robinson agrees that Newburyport is making slow, steady progress on its many initiatives, but acknowledges that it's not an easy task. "I'm a big believer in working from the bottom up," he says. "That's probably the most effective strategy. But what's really the most challenging is that nobody has time for this. There is so much going on in our lives and we are so busy it's hard to engage folks and get them to act. So, community outreach is probably the single biggest thing you have to work on in order to be effective."

Wildpoldsried, Germany

The Allgäu region of Germany is best known for its attractive rolling hills, forests, and lush pastures. The region supports numerous small farming villages where cattle and the dairy industry still play an important role in the local economy. But one of these villages, Wildpoldsried in southern Bavaria, is very different. While net-zero-energy targets are ambitious undertakings for most communities, Wildpoldsried has surpassed that goal by far and now produces over 300 percent more energy than it needs. With a population of only around twenty-six hundred, Wildpoldsried is generating almost $6 million (around four million euros) in annual revenue. This is all the more amazing for a farming community with no industry that just fifteen years earlier was trying to figure out how to attract new business, create new revenue, and build new municipal facilities—without going deeply into debt.

The agent of change for Wildpoldsried was a very conservative individual, Anton Zengerle, mayor of the municipality. But Zengerle had an open mind, and was willing to listen to others' suggestions—and not shy about offering his own. "Sustainability and conservative values are inextricably linked for me," he says.[9]

This all began in 1997 when the newly elected village council and mayor went for a weekend retreat to try to get their priorities sorted out. The main questions the council discussed were how they could better involve local citizens and how Wildpoldsried might look in 2020. Based on those early conversations, the mayor and council eventually came up with a more detailed mission statement in 1999 titled *Wildpoldsried Innovativ Richtungsweisend* (WIR) (Wildpoldsried Innovative Leadership), which would ultimately transform the sleepy little village into a renewable energy and sustainability powerhouse.

There were three main themes in the WIR statement: first, energy savings and renewable energy production; second, ecological construction of new buildings, mostly from wood; and third, water resource protection and ecological disposal of wastewater. The village council hoped that these broad guidelines would help inspire local citizens to do their part for the environment and create green jobs and businesses for the community and surrounding area. They succeeded beyond their wildest imaginings, and it took far less time to achieve most of their goals than they had originally thought.

Today Wildpoldsried is arguably the greenest village in Germany, and possibly in all of Europe. More than almost any other community, Wildpoldsried has managed to assemble a stunning array of locally owned renewable energy systems—including biogas, biomass, district heating, solar thermal, solar photovoltaics, hydropower, wind power, and geothermal—into a coordinated series of initiatives that more than offset the village's carbon footprint by 2012. Along the way, the village has managed to construct nine new community buildings complete with solar panels, an extensive district heating system with both municipal and private connections, ecological flood control, and a natural wastewater system—all without going deeply into debt.[10]

FIGURE 13-3. View of Wildpoldsried with solar panels on the roofs in the foreground and wind turbines on the hills in the distance. Photo by Susi Vogl.

This didn't happen overnight, of course. Even before the 1997 village council retreat, there were some early renewable energy and environmental initiatives. Zengerle has had a small twenty-five-kilowatt hydroelectric plant supplying electricity for his home since 1966. Two other local residents, Ludwig Schindele and Rainer Mayr, have had small hydropower plants for their sawmill and residence, respectively, since 1992.[11] And in 1994, the owner of a local landscape company constructed the first private natural wastewater system for his home, a project that helped inspire the village's far more ambitious natural wastewater project twelve years later.

One of the early pioneers to move forward with the renewable energy part of Wildpoldsried's plan was Wendelin Einsiedler, who, along with thirty of his neighbors, formed a civic society in 1999. The society organized a company that erected two Enercon E-58 wind turbines with a combined output of 3.5 megawatts on the hill behind the village in April 2000. The project was so successful that in 2001, the society formed a second company for two more turbines totaling 4.5 megawatts. A third turbine was added to this group in 2008. "The demand grew explosively," says Einsiedler.[12] Currently there are five wind turbines on the hill that, combined, produce about twelve million kWh of electricity annually. Wildpoldsried's total annual consumption of electricity is only 5.6 million kWh. Village residents were offered the chance to participate in the community wind projects, and around two hundred eventually invested various amounts, depending on what they could afford. "Initially there were reservations about the wind turbines," says Einsiedler. But the community was carefully informed about the projects, the issues were thoroughly discussed, and in the end the participants moved forward with confidence. The annual yield on the investments is from 8 to 10 percent, according to Einsiedler. It should come as no big surprise, then, that 86 percent of the locals say they are satisfied with the wind farms, according to one survey.[13]

Wendelin Einsiedler wasn't the only member of his family to embrace renewable energy. His brother, Ignaz, had already installed a biogas plant on his dairy farm back in 1997, and in 2000 he applied for municipal approval of a second plant. Two other farms in Wildpoldsried have biogas plants, one operated by Franz and Arthur Kolb, installed in 1996, and the other by Xaver Berkmiller and Christoph Schön, installed in 1999. Most of the electricity produced is used on the premises, with the excess fed into the grid. Excess heat produced by these plants is used to warm nearby homes.[14]

In addition to the biogas plants, at least five local residents have installed geothermal heating and cooling systems in their homes in recent years, which have resulted in substantial reductions in their energy bills. Since all of the electricity consumed in Wildpoldsried is generated by renewable sources, the electrically powered heat pumps are also 100 percent renewable. As more and more local citizens began to participate in one renewable energy project after another, the momentum really began to build. "It's not just about energy," says Zengerle. "It is also about the intelligent use of energy."[15]

The introduction of the national Renewable Energy Sources Act (Erneubare-Energien-Gesetz, EEG) in 2000, which granted priority to renewable energy sources, gave all renewable projects a boost—but especially solar installations in Wildpoldsried. One of the main features of this advanced renewable feed-in tariff is that it mandates long-term contracts from utilities for renewably produced electricity for twenty years based on the cost of generation. This makes the installation of relatively small-scale solar PV a moneymaking proposition for virtually everyone.

Consequently, after 2002, more and more solar panels were added to roofs in the village, in addition to those that had been installed as part of four earlier projects coordinated by the village planner, Thomas Knecht. In 2002, eighteen solar thermal (hot water) systems were installed in various locations. The following year, mostly photovoltaic panels totaling around sixty-seven kilowatts were added. That was followed in 2004 by many more PV panels, bringing the installed total to an impressive 865 kilowatts. In 2009, even more rooftops were covered with PV panels. In addition to numerous private roof-tops, municipal buildings that received PV panels included the fire station, city hall, maintenance yard, old gym, school, sports hall, and recycling center.[16]

But that's not all. Since Wildpoldsried is surrounded by extensive forest-land, there has been a long-standing initiative to replace the oil-fired heating systems in local buildings with wood-fired (pellet or chip) systems. In 1999, a wood-fired village district heating system was investigated but later post-poned due to high costs and lack of a suitable location for the central heating plant. However, after a good deal of additional research and planning, a wood-pellet-fired heating plant with a four-hundred-kilowatt thermal capacity was installed in the basement of a newly built community hall in 2005, while a system of distribution pipes was laid under the street to supply adjacent buildings with heat. The distribution system was extended in 2007, 2009, and again in 2010, and now serves forty-two public and private buildings.

Following the 1999 WIR mission statement's second theme, which focused on wood, Wildpoldsried's local forestry resources have also played a key role in the building materials used in many of the municipal buildings (and private homes) constructed in recent years. The showcase project is the Sports Hall, constructed over an eight-month period in 2004–05, with funding from the State of Bavaria. Most of the 10,370-square-foot building was built with wood. The south-facing roof was covered with a photovoltaic surface that produces 148 kilowatts of electricity. The building is surrounded by two soccer fields, a street hockey court, a beach volleyball court, a curling rink, and a firing range. Maintenance costs are covered entirely by club member-ship fees and income from the PV roof.[17]

The third theme in the 1999 WIR mission statement regarding water resource protection and the ecological disposal of wastewater was initially addressed in the fall of 2000 with the construction of an experimental water treatment facility funded by an EU grant obtained by Mayor Zengerle. The facility combined features similar to those found in conventional wastewater treatment facilities (settling tank, trickling plant filter, and clarifier for sludge removal) with a wetland known as WiWaLaMoor (Wildpoldsrieder Wasser-Landschaften im Moor). The ambitious goal was to send Wildpoldsried's effluent—after it was filtered and naturally cleaned—into the Leubas River as pure as the water already flowing there. An added challenge was the fact that the Leubas floods during the rainy season. A pilot project was completed in 2006, during which most of the kinks were worked out of the system; then a full-scale natural wetlands park was designed and built, complete with a trail and interpretive stations that explain the project to visitors.[18]

Wildpoldsried has also undertaken a wide range of energy-efficiency initiatives. In 2007, 250 conventional streetlight bulbs were replaced by energy-saving bulbs, resulting in a 30 percent savings in operating costs and a 40 percent improvement in the illumination. A home heating circulating pump exchange program for residents has resulted in substantial energy savings from the installation of more than 210 new high-efficiency pumps. Wildpoldsried also follows the Passive House program initiated by the federal government in 2008, and in 2011 the village council enacted an ordinance that requires all new home construction to follow similar guidelines. Among other requirements, these homes must be specially insulated and have energy-efficient appliances and no fossil fuel heating. The village also offers a thermal scanning initiative (including a partial rebate) that identifies areas of high heat loss in existing homes so they can be eliminated with energy-efficiency retrofits.[19]

Not surprisingly, Wildpoldsried has won numerous regional, national, and international awards for its many initiatives. In 2009, the village was awarded a Climate Protection Community award by German Environmental Aid, as well as the German Solar Prize from Eurosolar, the European Association for Renewable Energy.

As if that weren't enough, in April 2011, it was announced that Siemens and the regional utility, AÜW, would undertake a "Smart Grid of the Future" project in Wildpoldsried. The two-year project, a collaboration with the nearby Kempten University and others named IRENE (Integration of Renewable Energies and Electric Vehicles), will test new software and hardware to better coordinate the electricity network in combination with forty plug-in electric vehicles. Wildpoldsried was selected for the project because of its strong renewable energy portfolio.

In the past, tourism was virtually nonexistent in Wildpoldsried, but today, thanks to the village's international reputation as a renewable energy leader, a growing number of "energy tourists" are coming to see what is going on for themselves. "We are pleased that more and more international visitors are coming to us," says Susi Vogl, the village counsel coordinator. "We are happy to share our experiences and hope that many communities will embark on a similar path."[20] In 2012, a new Ecological Education Center opened, allowing groups of visitors to stay and participate in seminars and other energy and conservation-related events.

People who visit Wildpoldsried see a vision of the future. For many local residents, it's just the way things are—and they've already gotten used to it. In Mayor Zengerle's view, it's mostly become a matter of habit. "Our children don't know anything else," he says. "For them, the wind turbines belong on the mountain and the photovoltaic panels on the roofs of the village are just part of the picture."[21]

PART FOUR

A CALL
TO ACTION

Be Prepared

*We must take rapid, effective, innovative action to change
the ways we generate and use energy; renewable energy is ubiquitous,
offering a new model of energy generation that is local, democratic,
and free from the abuses of a centralized monopoly.*
—CALL TO ACTION FOR ENERGY DEMOCRACY,
RENEWABLE COMMUNITIES ALLIANCE

Be prepared. It's the Boy Scout motto. It's also good advice for the rest of us as we head deeper into the uncertainties of the twenty-first century, where many of our previous assumptions about a whole range of issues may no longer apply. The challenges we face with population, water, food, energy, and climate are enormous and interconnected in ways that make it hard to deal with one without creating problems elsewhere. It's easy to feel overwhelmed.

Adaptive Resilience

Rather than be paralyzed by these challenges, it's better to start dealing with them one small step at a time—as individuals and as communities. In his excellent chapter, "Personal Preparation," in the book *The Post Carbon Reader*, Chris Martenson suggests that there are many relatively inexpensive basic steps that anyone can undertake to better prepare for whatever may come. "The point of personal (and community) preparedness can be summed up in one single word: resilience." He goes on to say that we are more resilient when we have multiple (preferably local) sources and systems to supply our basic needs for food, water, heat, electricity, and so on. But he also admits that whatever we do will probably not be enough.

"Any steps we might take to prepare for a potential environmental, societal, or economic disruption, no matter how grand, are nearly certain to be insufficient. Nevertheless, they are still necessary," he says. "They will be insufficient because being perfectly prepared is infinitely expensive. But actions are necessary because they help us align our lives with what we know about the world. To put it all together, we take actions because we must. If we don't, who will? We change the world by changing ourselves."

Energy resilience is the thread that connects all of the strategies, initiatives, and projects we've looked at in this book. Greater energy resilience means that we are not hopelessly dependent on one (usually central) source of energy. We need alternatives. Even if you only cover a tiny percentage of your electricity use by adding a few photovoltaic panels connected to batteries to your home, you might have enough to get by during a power failure. "What's the difference between being zero percent self-reliant and three percent? Night and day," Martenson says. Heat your home with a renewable fuel, and you are instantly more resilient in the face of rising fossil fuel prices. Scale that up to the community level, and you might have enough local power to keep essential services running for thousands of people.

We have seen that, unlike the fossil-fuel-based centralized power systems of the twentieth century, renewable energy in the twenty-first century can be scaled up or down to meet local needs. We've also seen that the centralized, corporation-dominated fossil fuel industry requires huge concentrations of capital and expensive infrastructure, while local, distributed renewable energy is relatively affordable and available almost anywhere without a lot of that supporting infrastructure. In fact, wind, solar, and geothermal are now so commercially viable that thirty-one states could meet their entire electricity needs with in-state renewable energy resources according to one recent study.[1] Lack of renewable energy sources, or the technology to harness them, is not a problem.

Challenges

Nevertheless, renewable energy developers large and small face quite a few challenges. The lack of a coherent national energy policy is right at the top of the list. Ongoing resistance from many utilities is another. The complexity of financing community energy projects is yet another major challenge, but with careful research and informed financial and legal advice, most sound projects should be able to attract the equity and debt financing they need. Inflexible and outdated federal, state, and local regulatory and approval processes are another major barrier, although some progress has been made in recent years. "Permitting is often the biggest challenge for our customers," says Mike New, vice president of Canyon Hydro, a hydroelectric turbine manufacturer in Deming, Washington.[2] And, as we have seen over and over again, inconsistent government incentives and other supports for the renewable energy sector are a huge problem for anyone trying to put together a prudent, long-range business plan for almost any project. One year, federal supports are available. The next, Congress changes its mind, and the supports are gone. This has happened repeatedly for decades, and it drives almost everyone in the renewable energy sector crazy. Contacting your state and federal representatives and urging them to enact long-term support for renewables is one of the most important things that you can do to promote energy resilience.

Another frequent challenge to local energy is public opposition—the "not in my backyard" phenomenon. This applies to virtually all renewable energy projects. "There is usually public resistance to any type of new power project, and for good reason," Mike New says. "Every form of electricity production, no matter how green, has a negative impact on the environment. It's important to help people understand where the electricity they use comes from, and to show why a new source of renewable power is actually better for the overall environment, not worse." People simply have to understand that all energy projects, including renewables, involve trade-offs. The real question is whether the potential benefits outweigh the negative impacts. With most renewables, they do, especially if the project is intelligently sized and sited to minimize impacts on nearby residents.

The alternative, business as usual, no longer makes economic sense. A 2010 study by the Civil Society Institute evaluated a strategy for the U.S. electric industry that provided large-scale public health and environmental benefits at a reasonable cost. Updated in 2011 to reflect recent decreases in the costs of renewables, the study compared "business as usual" (BAU) to a "Transition Scenario" in which the nation moves toward a power system based on efficiency and renewable energy. Under this new scenario, all coal-fired power plants are retired, along with nearly a quarter of the nation's nuclear power plants, by 2050. Unlike BAU, the Transition Scenario does not rely on hoped-for breakthroughs: Nearly all demand is projected to be met with technologies that are commercially available today. Most remarkable of all, the Transition Scenario is likely to be less expensive than BAU. "The idea that we could capture the kind of benefits this scenario provides while also saving money is a significant change in our thinking about this industry," the study says. "It reflects a fundamental shift in the cost of renewable energy relative to fossil-fueled and nuclear energy."[3] While these are the conclusions of only one study, it's clear that some of our long-standing assumptions about the relative costs of conventional energy versus renewables are changing. This offers a lot of potential opportunity for local power advocates who focus on renewables and energy efficiency.

Opportunities and Possibilities

Renewable energy unquestionably lends itself to a decentralized system of power generation and ownership, and also increases local energy resilience. This represents an enormous opportunity for local energy projects. "There is an incredible number of communities in our nation that can be carbon-neutral based on their own local renewable resources," says Terry Meyer, the owner of Cascade Community Wind Co. in Bellingham, Washington. "I don't think most people realize how much renewable energy is out there."[4]

While renewable energy resources are available almost everywhere, they are unevenly distributed across the nation. Biomass, for example, is especially abundant in the Northeast, Upper Midwest, and Northwest, while solar potential is strongest in the Southwest. Communities in these regions will especially benefit by taking advantage of the ample renewable resources that are available to them. But communities anywhere still have plenty of opportunity to revitalize their local economies with local power projects.

We've seen that keeping your energy dollars circulating in the local economy has many positive effects, especially if renewable energy projects are locally owned. Local community ownership is the key. Instead of sending money out of state (or out of country) to some distant corporate entity, dollars spent on local energy projects have a multiplier effect. Local renewable energy projects can provide short-term jobs during the construction phase, long-term employment for system operation and maintenance personnel, and long-term certainty about supply and price. "With renewables, you are investing in something that hedges your long-term price," says Mike Antheil, the executive director of the Florida Alliance for Renewable Energy. "Fossil fuels are depleting commodities, and their prices will generally be going up over the long term. But the price of most renewables is predictable and safe, and the cost of the technology will continue to drop over time."[5]

We've also seen that the collaborative effort required for most local energy projects tends to strengthen the community in many ways. With every project, local knowledge and expertise in renewables increase. As community members collaborate on these initiatives, they develop new skills and relationships that provide the experience and confidence for additional, larger projects—even projects that extend beyond local energy to help promote the broader relocalization of the economy. More sustainable and self-sufficient communities are the result.

Having said all of this, it's important to remember that none of the individual strategies we have covered in this book is a "silver bullet" solution that will replace our current demand for energy. That's why it's so important to consider *all* of the renewable strategies along with conservation and energy efficiency that are available and appropriate for your community.

Take the First Step

If you are working toward greater energy resilience, you are not alone. This is a group effort, and the more people who get involved, the better. There has been a strong emphasis on community in this book. That's because community is a key part of resilience. This is a lesson that was driven home powerfully in the aftermath of Hurricane Irene, which blew through my home state of Vermont in August 2011; the storm flooded nearly every stream and river in the state, destroying numerous homes, wrecking five hundred miles of highways, and damaging or destroying two hundred bridges. The quick response of Vermonters helping one another at the community level was truly inspiring and provides a great model for others to follow in future calamities. You are only as secure as your neighbors. And you and your neighbors are only as secure as your community.

A good way to get started is to see if there is a Transition Initiative or other local sustainability group near you. If not, start one with some like-minded people who also want to help your community become more resilient. At first, it may just be two or three people with whom you share values and whose company you enjoy. But soon you will attract more enthusiastic people who want to help. And we shouldn't lose sight of maintaining a positive, optimistic attitude about our efforts. It can actually be fun, despite all the hard work and the serious nature of the challenges ahead.

Community collaboration is useful in another way. Involving the community from the very beginning in any local renewable energy project is extremely important in helping to build support and also in avoiding opposition caused by misinformation. "None of us is smarter than all of us," says Jon Folkedahl, who helped organize the community wind project in Willmar, Minnesota. "There is always the potential to overlook crucial factors in any project like this, so the more people you can get thinking and talking about it and offering opinions and advice, the less likely you are to make mistakes."[6] A long list of successful community energy projects across the nation proves that *the whole is truly greater than the sum of its parts.*

Working on a community energy project also helps build a stronger sense of community, according to Joy Hughes, founder and CEO of Solar Panel Hosting in Westminster, Colorado. "The greatest thing about it is the spirit that we see in the community when people get together and realize that projects like this can work for them," she says. "It's more than just a solar project; it strengthens the sense of community. They may have talked about sustainability in the past, but this actually has them working together to achieve it. That's so incredibly powerful."[7]

Admittedly, working with a large number of people takes time and patience, and organizing a community renewable energy project can be a slow process. But it can be enormously rewarding. "Most of these projects take a while," says Lynn Benander, the CEO of Co-op Power and Northeast Biodiesel in Greenfield, Massachusetts. "Community-based business does take time, and you need to have people who are willing to go in with their eyes wide open. But the most important factor for success is determination and commitment. If a group has that, and is willing to go the long route, you'll have an amazing asset that can stay in your community for generations to come. So in my mind it's well worth it."[8]

Still, we can't ignore the reality of the forces we have unleashed with our profligate burning of fossil fuels. Global warming is accelerating rapidly, surpassing most previous scientific predictions. This requires a massive, coordinated international response if we are to at least mitigate some of the worst effects. Yet at the international level, most political leaders continue to dither and stall while glaciers around the world melt, sea levels rise faster than projected, and global climate and weather patterns become increasingly unstable and dangerous.

Lyle Estill, president of Piedmont Biofuels in Pittsboro, North Carolina, recognizes how serious a threat global warming is to human society. "My father served in World War II. He went off to war because it was the moral imperative of their day; it wasn't even questioned," he says. "I think that climate change is the moral imperative of our time. And if I live long enough to have grandchildren climb onto my lap, I want to be able to tell them what I did to combat climate change. I want to be able to tell them that we threw everything we had available at it in order to fight for the moral imperative of our time."[9]

The stakes are high, but the goal of a livable future for our children and grandchildren is definitely worth the effort. Local energy is an important part of this new imperative. Get your own house in order by becoming as energy-resilient as you can, and then join with your neighbors to relocalize the economy while increasing your community's resilience at the same time. So much to do. So little time. Better get to work.

RESOURCES

Find more resources at the companion website to this book, resilience.org.

General Information

American Public Power Association
Washington, DC
www.appanet.org
APPA is a service organization for the nation's more than two thousand community-owned electric utilities, which serve more than forty-three million Americans.

Database of State Incentives
for Renewables & Efficiency
www.dsireusa.org
DSIRE is a comprehensive source of information on state, local, utility, and federal incentives and policies that promote renewable energy and energy efficiency.

Earth Advantage Institute
Portland, OR
www.earthadvantage.org
EAI works with the building industry to help implement sustainable building practices.

Earth Policy Institute
Washington, DC
www.earth-policy.org
EPI, headed by acclaimed environmental analyst Lester Brown, is dedicated to providing a vision of an environmentally sustainable economy as well as a road map of how to get from here to there.

Florida Alliance for Renewable Energy
www.floridaallianceforrenewableenergy.org
FARE is dedicated to creating jobs and economic benefit for the citizens of Florida by the use of renewable energy through effective public policy.

Home Power magazine
Ashland, OR
www.homepower.com
This informative magazine is a wonderful resource for small-scale renewable energy strategies in general.

Institute for Local Self-Reliance
Washington, DC
www.ilsr.org
ILSR promotes sustainable, self-reliant communities.

Midwest Renewable Energy Association
Custer, WI
www.midwestrenew.org
MREA promotes renewable energy, energy efficiency, and sustainable living through education and demonstration. The site offers information about events, education and training, discussion forums, and much more.

Minnesotans for Sustainability
www.mnforsustain.org
Although its name includes *Minnesota*, this is one of the best and most comprehensive sustainability websites anywhere.

Mother Earth News magazine
Topeka, KS
www.motherearthnews.com
Mother Earth News is the original guide to living wisely, with articles on green building, alternative energy, organic gardening, natural health, homesteading, whole foods, transportation, and more.

National Religious Partnership
for the Environment
Amherst, MA
www.nrpe.org
NRPE is an association of independent faith groups across a broad spectrum working for environmental sustainability and justice.

National Renewable Energy Laboratory
Golden, CO
www.nrel.gov
NREL's comprehensive site has information on solar, wind, biomass, geothermal, and much more.

Northeast Organic Farming Association
Stevenson, CT
www.nofa.org
NOFA is an organization of nearly
four thousand farmers, gardeners,
and consumers working to promote
healthy local food, organic farming
practices, and a cleaner environment.

Office of Energy Efficiency
and Renewable Energy
U.S. Department of Energy
www.eere.energy.gov
EERE is an excellent government source
for renewable energy information.

Rocky Mountain Institute
Snowmass, CO
www.rmi.org
RMI is an independent, entrepreneurial,
nonprofit think-and-do tank fostering the
efficient and restorative use of resources to
create a more sustainable world.

Transition US
www.transitionus.org
Transition US provides inspiration,
support, networking, and training
for Transition initiatives across the
United States. It works in close
partnership with the Transition
Network, a UK-based organization
that supports the international
Transition Movement.

U.S. Green Building Council
www.usgbc.org
USGBC developed the LEED rating system
in 2000 that provides building owners and
operators a framework for identifying and
implementing measurable green building
design, construction, operations, and
maintenance solutions.

Building Community

Community Solutions
Yellow Springs, OH
www.communitysolution.org
CS provides knowledge and
practices for low-energy living
and self-reliant communities.

Ecocity Builders
Oakland, CA
www.ecocitybuilders.org
Ecocity Builders develops and implements
policy, design, and educational tools and
strategies to build thriving urban centers
and to reverse patterns of sprawl and
excessive consumption.

Island Institute
Rockland, ME
www.islandinstitute.org
The Island Institute helps coastal
Maine communities remain vibrant
places to live and work through a variety
of programs and services, including
assistance with local energy projects.

Community Energy

Acorn Renewable Energy Co-op
Middlebury, VT
www.acornenergycoop.com
Acorn is a regional community renewable
energy and efficiency cooperative.

The Community Energy Exchange
Burlington, VT
www.communityenergyexchange.com
The exchange is developing an online platform through which community members
will be able to directly finance clean energy
projects installed in their region.

Community Energy Partners
Freeport, ME
www.communityenergypartners.com
Comm-En supports community energy development as a local solution to global problems.

Community Energy Partnerships Program
Toronto, Ontario, Canada
www.communityenergyprogram.ca
CEPP funds community power
projects and renewable energy
education projects in Ontario.

Community-Based Energy Development
Minneapolis, MN
www.c-bed.org
C-BED fosters, promotes, and secures the local
economic development and environmental
benefits attached to renewable energy
production facilities that are owned by
ordinary members of local communities.

Co-op Power
Greenfield, MA
www.cooppower.coop
Co-op Power is a regional community renewable energy and efficiency cooperative that is also the developer of Northeast Biodiesel Company LLC.

Local Clean Energy Alliance
Oakland, CA
www.localcleanenergy.org
The alliance is a coalition of nonprofits, businesses, and community groups working to advance strong climate protection plans and create thousands of clean energy jobs throughout the San Francisco Bay Area.

Northwest Community Energy
Northwest Sustainable Energy for Economic Development
Seattle, WA
www.nwcommunityenergy.org
This site is the community energy hub of Northwest SEED, which works directly with communities to build sustainable local economies founded on locally owned energy.

Ontario Sustainable Energy Association
Toronto, Ontario, Canada
www.ontario-sea.org
OSEA is a provincial, nonprofit umbrella organization formed to implement community sustainable energy projects across Ontario. This site lists workshops, publications, and other educational materials.

Plymouth Area Renewable Energy Initiative
Plymouth, NH
www.plymouthenergy.org
PAREI is a regional community renewable energy and efficiency organization.

Transportation

The Bike-sharing Blog
bike-sharing.blogspot.com
Provides information on the emerging public transport mode of bike sharing.

CarSharing.net
www.carsharing.net
This site offers a good introduction to and overview of car sharing as well as links to programs in North America and around the world.

Peak Energy

There are hundreds (if not thousands) of peak oil/energy sites on the Internet. Here are a few of the better ones.

Association for the Study of Peak Oil & Gas—USA
Washington, DC
www.aspo-usa.org
ASPO-USA is a nonpartisan research and public education initiative to address America's peak oil energy challenge.

Resilience.org
The energy section of resilience.org (formerly energybulletin.net) is simply one of the best energy sites on the Internet. It's updated daily with high-quality articles and commentary from around the world on all types of energy. An outstanding resource.

Post Carbon Institute
Santa Rosa, CA
www.postcarbon.org
PCI provides individuals, communities, businesses, and governments with the resources needed to understand and respond to the interrelated economic, energy, environmental, and equity crises that define the twenty-first century.

The Oil Drum
www.theoildrum.com
An excellent forum for the discussion of peak oil and other energy issues moderated by "a bunch of academics and other folks who think they know what they are talking about." They do.

Willits Economic Localization
Willits, CA
well95490.org
One of the oldest and most successful relocalization response initiatives in the United States.

Household Energy Resilience

Community-Based Programs

Cambridge Energy Alliance
Cambridge, MA
www.cambridgeenergyalliance.org
CEA works with homeowners, businesses, and institutions to achieve high levels of energy savings and increase Cambridge's use of clean energy sources.

CharlestonWISE
Charleston, SC
www.charlestonwise.com
Operated by the Sustainability Institute, this program helps people arrange a home energy assessment, decide what improvements make sense, choose a contractor, and coordinate rebates and tax credits.

City of Berkeley Home Energy
Upgrade Programs
Berkeley, CA
www.ci.berkeley.ca.us/ContentDisplay
.aspx?id=19376
The City of Berkeley offers a wide range of energy-efficiency programs, including home performance testing, home energy upgrades, and much more.

City of Palm Springs, CA
www.palmspringsca.gov/index
.aspx?page=772
The City of Palm Springs offers energy-efficiency and energy reduction programs for businesses and residents.

Clean Energy Coalition
Ann Arbor, MI
www.cec-mi.org
Clean Energy Coalition promotes clean energy technologies as a way to create healthier energy-independent communities.

Community Energy Challenge
Bellingham, WA
www.communityenergychallenge.org
The Community Energy Challenge provides participating households and businesses throughout Whatcom County with energy assessments; customized energy action plans; and assistance with rebates, financing opportunities, and other services.

Energize Bedford
Mount Kisco, NY
www.energizebedford.org
Energize Bedford is a community-based energy-efficiency program.

Greater Cincinnati Energy Alliance
Cincinnati, OH
www.greatercea.org
The Energy Alliance provides reduced-cost energy assessments and financial incentives to support home energy improvements.

Pratt Center for Community Development
Brooklyn, NY
www.prattcenter.net
The Pratt Center for Community Development brings together professionals, educators, and graduate students in architecture, urban planning, and related fields to collaborate with community-based partners to build sustainable and successful city neighborhoods.

San Francisco Energy Efficiency
San Francisco, CA
www.sfenvironment.org/our_programs
/topics.html?ssi=1&ti=14
The City of San Francisco has a wide range of energy-efficiency and energy reduction programs for both residents and businesses.

The Energy Coalition
Irvine, CA
www.energycoalition.org
TEC mobilizes, educates, and empowers communities to change the way they use and think about energy.

State-Based Programs
Clean Energy Works Oregon
www.cleanenergyworksoregon.org
Clean Energy Works Oregon is a public–private alliance among Energy Trust of Oregon, utilities, financial institutions, local communities, and contractors.

Connecticut Energy Efficiency Fund
www.ctsavesenergy.org
CEEF promotes energy efficiency and helps residents and businesses get in the habit of using energy more efficiently.

Efficiency Maine
Augusta, ME
www.efficiencymaine.com
Efficiency Maine helps businesses and residents use energy resources more efficiently, reduce energy costs, and lighten the impact on the state's environment from the burning of fossil fuels.

Energy Trust of Oregon
Portland, OR
www.energytrust.org
The Energy Trust is an independent organization helping utility customers benefit from saving energy and generating renewable energy.

Efficiency Vermont
Burlington, VT
www.efficiencyvermont.com
Efficiency Vermont provides technical assistance, rebates, and other financial incentives to help households and businesses reduce their energy costs with energy-efficient equipment, lighting, and approaches to construction and major renovation.

Flex Your Power
San Francisco, CA
www.fypower.org
Flex Your Power is California's statewide energy-efficiency marketing and outreach campaign.

Minnesota Energy Challenge
Minneapolis, MN
www.mnenergychallenge.org
The Energy Challenge provides in-depth guides to energy-efficiency and conservation actions, as well as practical information about savings and cost.

New York State Energy Research and Development Authority
Albany, NY
www.getenergysmart.org
NYSERDA helps homeowners and renters as well as apartment and other building owners reduce their energy costs.

Rhode Island Office of Energy Resources
www.energy.ri.gov
The office offers programs to assist residents, businesses, cities and towns, and institutions with their energy needs.

Wisconsin Energy Conservation Corporation
Madison, WI
www.weccusa.org
WECC is a national leader in the design and implementation of award-winning energy savings programs for the residential and business sectors, utilities, municipalities, and regulators across the United States and Canada.

Federal Programs

Energy Efficient Mortgage Program, Federal Housing Administration
portal.hud.gov/hudportal/HUD
?src=/program_offices/housing/sfh/eem/energy-r
The EEM helps homebuyers and homeowners save money on utility bills by enabling them to finance the cost of adding energy-efficiency features to new or existing housing as part of their FHA-insured home purchase or mortgage refinancing.

Energy Star Products Tax Incentives
www.energystar.gov/taxcredits
This site is a good source for up-to-date information on energy-related financial incentives.

Tax Incentives Assistance Project
www.energytaxincentives.org
TIAP helps consumers and businesses make use of the federal income tax incentives for energy-efficient products and technologies passed by Congress as part of the Energy Policy Act of 2005 and subsequently amended several times.

Weatherization Assistance Program
www1.eere.energy.gov/wip/wap.html
WAP helps low-income families reduce their energy bills by making their homes more energy-efficient.

Energy-Specific Resources

Biogas

Farm Power Northwest LLC
Skagit County, WA
www.farmpower.com
Farm Power Northwest is a developer of small-scale, farm-based manure digesters that generate electricity and provide fertilizer and bedding for participating farms.

Biomass

25 x '25
www.25x25.org
25 x '25 is an agriculturally based initiative aimed at producing 25 percent of America's energy from renewable resources by 2025.

Biomass Energy Foundation
biomassenergyfndn.org
BEF publishes of a wide range of books on all aspects of biomass energy, but especially on gasification for heat, power, and fuel.

Biomass Energy Resource Center
Montpelier, VT
www.biomasscenter.org

BERC works on projects around the country to install systems that use biomass fuel to produce heat and/or electricity.

HearthNet
www.hearth.com
This site is a comprehensive source of information on wood-, pellet-, coal-, and gas-burning hearth appliances and central heaters with links to manufacturers and local retailers.

Northern Forest Center
Concord, NH
www.northernforest.org
The center is creating a new generation of conservation tools that can provide economic incentives for good long-term forest stewardship while reinvigorating the economy.

Pellet Fuels Institute
Arlington, VA
www.pelletheat.org
PFI is a trade association that represents the fuel preparation and clean-burning technology of renewable biomass energy resources.

The Wood Heat Organization
www.woodheat.org
The Wood Heat Organization offers detailed information on wood-burning technologies, chimneys, firewood, safety, environmental issues, questions and answers, links, and much more.

Geothermal
Geo-Heat Center, Oregon Institute of Technology
Klamath Falls, OR
geoheat.oit.edu
The center provides information developed through firsthand experience with hundreds of projects and through extensive research.

Geothermal Energy Association
Washington, DC
www.geo-energy.org
GEA is a trade organization comprising U.S. companies that support the expanded use of geothermal energy.

Geothermal Exchange Organization (GEO)
Washington, DC

www.geoexchange.org
GEO is a trade association for the geothermal heat pump industry in the United States.

Geothermal Resources Council
Davis, CA
www.geothermal.org
GRC is the primary professional educational association for the international geothermal community.

International Ground Source Heat Pump Association
Oklahoma State University
Stillwater, OK
www.igshpa.okstate.edu
IGSHPA advances geothermal heat pump technology on the local, state, national, and international levels.

Hydro
Canyon Hydro
Deming, WA
www.canyonhydro.com
Manufacturer of hydroelectric system components.

Energy Systems & Design
Sussex, New Brunswick, Canada
www.microhydropower.com
Canadian manufacturer of micro-hydro systems.

Federal Energy Regulatory Commission
Small/Low-Impact Hydropower Program
www.ferc.gov/industries/hydropower/gen-info/licensing/small-low-impact.asp
This website explains how to obtain FERC authorization to construct and operate small/low-impact projects.

Low Impact Hydropower Institute
Portland, ME
www.lowimpacthydro.org
LIHI aims to reduce the impacts of hydropower dams through market incentives, primarily through the certification of environmentally responsible, low-impact hydropower.

Microhydropower.net
An informative Internet portal on micro-hydropower in nations around the world.

National Hydropower Association
Washington, DC
www.hydro.org
NHA is a national trade association
for the hydropower industry.

Ocean Energy Council, Inc.
West Palm Beach, FL
www.oceanenergycouncil.com
OEC educates about and advocates
for ocean energy (tidal, current, wave,
thermal, and wind).

Ocean News & Technology magazine
Palm City, FL
www.ocean-news.com
The only publication of its kind,
reporting on the latest ocean
industry news, events, and technology
developments around the world.

Liquid Biofuels

Bio Diesel America
www.biodieselamerica.org
A comprehensive site about
everything biodiesel, including where
to buy it, forums, and much more.

Canadian Renewable Fuels Association
Toronto, Ontario, Canada
www.greenfuels.org
CRFA promotes ethanol and biodiesel.

Collaborative Biodiesel Tutorial
biodieseltutorial.utahbiodieselsupply.com
This site will help you learn how to
make your own biodiesel safely, with
instructions and advice from experienced
producers around the world.

Journey to Forever
www.journeytoforever.org
A highly educational website containing
a wealth of information about ethanol,
biodiesel, and many other sustainable
living subjects, including links to additional
sources (in English, Japanese, and Chinese).

National Biodiesel Board
Jefferson City, MO
www.biodiesel.org
An excellent and extensive source of
current industry information on biodiesel,
with news, technical information, links,
and much more.

The Online Distillery Network
for Distilleries and Fuel Ethanol
Plants Worldwide
www.distill.com
Provides links to the home pages
of distilleries and fuel-ethanol plants
located around the world, and to related
organizations and other services.

Renewable Fuels Association
Washington, DC
www.ethanolrfa.org
The U.S. national trade association for the
ethanol industry, including links, news,
technical reports, and general information.

Solar

American Solar Energy Society
Boulder, CO
www.ases.org
ASES publishes *Solar Today* magazine.

Cooperative Community Energy
San Rafael, CA
www.ccenergy.com
A buyers' cooperative providing quality
solar energy systems to its members at
reasonable prices.

Florida Solar Energy Center
Cocoa, FL
www.fsec.ucf.edu
Provides highly regarded independent
third-party testing and certification of
solar hot water systems and other solar or
energy-efficient products.

Solar Energy International
Carbondale, CO
www.solarenergy.org
SEI provides education and technical
assistance so that others will be
empowered to use renewable energy
technologies around the world.

Solar Gardens Institute
Westminster, CO
www.solargardens.org
The Solar Gardens Institute provides
tools and training to co-develop
community solar arrays.

Solar Living Institute
Hopland, CA
www.solarliving.org

SLI offers workshops on a wide variety of renewable energy, energy conservation, and building technologies.

Surface Meteorology and Solar Energy, National Aeronautics and Space Administration
eosweb.larc.nasa.gov/sse
Complete solar energy data for the entire planet, compliments of NASA.

Wind

Cascade Community Wind Company
Bellingham, WA
cascadecommunitywind.com
A community wind project developer.

Hull Wind
Hull, MA
www.hullwind.org
Home site for the Hull I and Hull II community-owned wind turbines in Hull, Massachusetts.

Wind-Works
Tehachapi, CA
www.wind-works.org
An excellent online archive of articles and commentary, primarily but not solely on wind energy. Maintained by acclaimed wind energy expert Paul Gipe.

WindShare
Toronto, Ontario, Canada
www.windshare.ca
Home site for the WindShare community-owned turbine in Toronto, Ontario.

Windustry
Great Plains Windustry Project
Minneapolis, MN
www.windustry.com
Simply one of the best and most informative wind power sites, promoting wind energy through outreach, technical assistance, and nonprofit collaborations.

Biodiesel Cooperatives

The Berkeley Biodiesel Collective
Berkeley, CA
www.berkeleybiodiesel.org

Boulder Biodiesel Cooperative
Boulder, CO
www.boulderbiodiesel.com

Everpure Biodiesel Co-op
Erin, Ontario, Canada
www.everpurebiod.ca

Grease Works! Biodiesel Cooperative
Corvallis, OR
www.grease-works.com

Louisville Biodiesel Cooperative
Louisville, KY
www.louisvillebiodieselcoop.org

Piedmont Biofuels
Pittsboro, NC
www.biofuels.coop

ENDNOTES

Introduction

1. "At Least 24 Killed as Tornado Shreds Joplin," *Kansas City Star*, May 22, 2011, www.kansascity.com/2011/05/22/2894920/tornado-shreds-joplin.html.
2. Mary Williams Walsh, "Irene Adds to a Bad Year for Insurance Industry," *New York Times*, August 28, 2011, www.nytimes.com/2011/08/29/business/irene-damage-may-hit-7-billion-adding-to-insurer-woes.html?pagewanted=all.
3. "State of the Climate, Wildfires," National Oceanic and Atmospheric Administration, National Climate Data Center, December 2011, updated January 6, 2012, www.ncdc.noaa.gov/sotc/fire.
4. "Global Warming Fast Facts," National Geographic News, http://news.nationalgeographic.com/news/2004/12/1206_041206_global_warming.html, accessed January 30, 2012; "Global Warming Is Accelerating," National Wildlife Federation, www.nwf.org/Global-Warming/What-is-Global-Warming/Global-Warming-is-Accelerating.aspx, accessed January 30, 2012.
5. National Oceanic and Atmospheric Administration, "NOAA: 2010 Tied for Warmest Year on Record," January 12, 2011, www.noaanews.noaa.gov/stories2011/20110112_globalstats.html.
6. "2010 a Deadly Year for Natural Disasters," Associated Press, December 19, 2010.
7. Named after the Shell geologist Dr. Marion King Hubbert who, in 1956, accurately predicted that U.S. domestic oil production would reach its peak in 1970.
8. International Energy Agency, "World Energy Outlook 2010, Executive Summary," www.worldenergyoutlook.org/docs/weo2010/WEO2010_ES_English.pdf.

Chapter 1

1. SolarFest has since moved to a new location at Forget-Me-Not Farm in Tinmouth, Vermont.
2. Ten of the eleven recessions since World War II were preceded by an oil price spike. J. D. Hamilton, "Causes and Consequences of the Oil Shock of 2007–08," *Brookings Papers on Economic Activity*, Spring 2009, pp. 215–59.
3. U.S. Energy Information Administration, "Demand," www.eia.gov/pub/oil_gas/petroleum/analysis_publications/oil_market_basics/demand_text.htm.
4. Oil Market Report, "Highlights of the Latest OMR," International Energy Agency, June 16, 2011, http://omrpublic.iea.org.
5. Michael Winter, "Ohio Quakes Linked to Oil-Drilling Waste Pumped into Wells," *USA Today*, January 2, 2012, http://content.usatoday.com/communities/ondeadline/post/2012/01/11-ohio-quakes-linked-to-oil-drilling-waste-pumped-in-wells/1.
6. Ian Urbina, "Insiders Sound an Alarm Amid a Natural Gas Rush," *New York Times*, June 25, 2011, www.nytimes.com/2011/06/26/us/26gas.html. See also David Hughes, "Will Natural Gas Fuel America in the 21st Century?," Post Carbon Institute, 2011, www.postcarbon.org/naturalgas.
7. James Hansen, "Coal-Fired Power Stations Are Death Factories. Close Them," *Observer*, February 14, 2009, www.guardian.co.uk/commentisfree/2009/feb/15/james-hansen-power-plants-coal.
8. Keith Bradsher, "China Outpaces U.S. in Cleaner Coal-Fired Plants," *New York Times*, May 10, 2009, www.nytimes.com/2009/05/11/world/asia/11coal.html.
9. Patrick Reis, "Study: World's 'Peak Coal' Moment Has Arrived," *New York Times*, September 29, 2010, www.nytimes.com/gwire/2010/09/29/29greenwire-study-worlds-peak-coal-moment-has-arrived-70121.html.

10. Richard Heinberg, *Blackout: Coal, Climate and the Last Energy Crisis* (Gabriola Island, BC: New Society, 2009).
11. Gloria Shur Bilchik, "No Good News About Nuclear Waste, Says GAO," Occasional Planet, June 14, 2011, www.occasionalplanet.org/2011/06/14/no-good-news-about-nuclear-waste-says-gao-2.
12. Jonathan Fahey and Ray Henry, "U.S. Nuclear Waste Increasing with No Permanent Storage Available," Huffpost Green, March 22, 2011, www.huffingtonpost.com/2011/03/23/us-nuclear-waste-radioactive-storage_n_839438.html.
13. Paul Blumenthal, "Nuclear Industry Lobbyists Battle Fallout from Japan Reactor Crisis," blog, Sunlight Foundation, March 23, 2011, http://sunlightfoundation.com/blog/2011/03/23/nuclear-industry-lobbyists-battle-fallout-from-japan-reactor-crisis.
14. Joseph Romm, "The Staggering Cost of New Nuclear Power," Center for American Progress, January 5, 2009, www.americanprogress.org/issues/2009/01/nuclear_power.html.
15. Jeff Donn, "Safety Standards Lowered for Aging Nuclear Reactors," Associated Press, June 20, 2011, www.burlingtonfreepress.com/apps/pbcs.dll/article?AID=2011110620012.
16. Jeff Donn, "Radioactive Tritium Leaks Found at 48 U.S. Nuke Sites," Associated Press, June 21, 2011, www.msnbc.msn.com/id/43475479/ns/us_news-environment.
17. Agustino Fontevecchia, "Solar Industry Rising: Can Energy's Fastest-Growing Sector Keep It Going?" *Forbes*, March 25, 2011, http://blogs.forbes.com/afontevecchia/2011/03/25/solar-industry-rising-can-energys-fastest-growing-sector-keep-it-going/?utm_medium=twitter.
18. David Appleyard, "From the Editor: Wind Sector Still Growing Fast," RenewableEnergyWorld.com, May 31, 2011, www.renewableenergyworld.com/rea/news/article/2011/05/from-the-editor15.
19. Bridgette Meinhold, "Norway to Build World's Largest Wind Turbine," inhabitat, February 16, 2010, http://inhabitat.com/norway-to-build-the-worlds-largest-wind-turbine.
20. Timon Singh, "Spain to Build World's Largest Wind Turbine," inhabitat, December 2, 2010, http://inhabitat.com/spain-to-build-worlds-largest-wind-turbine.
21. Russell Ray, "Rediscovering Hydropower," blog, RenewableEnergyWorld.Com, November 22, 2010, www.renewableenergyworld.com/rea/blog/post/2010/11/rediscovering-hydropower.
22. Elton Robinson, "USDA Gets It Right by Changing Corn for Ethanol Category," Farm Press Blog, April 20, 2011, http://deltafarmpress.com/corn/usda-gets-it-right-changing-corn-ethanol-category.
23. The key points for this section on structural challenges are synthesized from David Fridley, "Nine Challenges of Alternative Energy," in *The Post Carbon Reader: Managing the 21st Century's Sustainability Crises*, Richard Heinberg and Daniel Lerch, eds. (Healdsburg, CA: Watershed Media, 2010).
24. David Murphy. "A Tour of the Energy Terrain," in *Energy: Overdevelopment and the Delusion of Endless Growth*, eds. Tom Butler and George Wuerthner (Healdsburg, CA: Watershed Media, forthcoming).
25. Munich Re, press release, "Two Months to Cancún Climate Summit/Large Number of Weather Extremes as Strong Indication of Climate Change," September 27, 2010, www.munichre.com/en/media_relations/press_releases/2010/2010_09_27_press_release.aspx.

Chapter 2

1. "Smoke, Mirrors & Hot Air: How ExxonMobil Uses Big Tobacco's Tactics to Manufacture Uncertainty on Climate Science," Union of Concerned Scientists, January 2007, www.ucsusa.org/assets/documents/global_warming/exxon_report.pdf.
2. Con Edison Co. of NY, Inc., "Con Edison Initiates Energy Demand Reduction Programs in Lower Manhattan, Middle of Staten Island," *Transmission & Distribution World*, March 6, 2008, http://tdworld.com/substations/conedison-substation-building.
3. BGE, "BGE Customers Continue to Strongly Support Energy Savers Program," *Transmission & Distribution World*, March 15, 2011, http://tdworld.com/customer_service/bge-smart-savers-0311/index.html.

4. Robert Lamb, "Is The Smart Grid a Dumb Idea?," Discovery News, June 22, 2010, http://news.discovery.com/tech/is-the-smart-grid-a-dumb-idea.html.
5. Associated Press, "Experts Warned of Weak Power Grid," *Wired*, August 15, 2003, www.wired.com/science/discoveries/news/2003/08/60057.
6. "Pickens Says U.S. Spent $337 Billion on Oil Imports," Bloomberg, January 19, 2011, www.bloomberg.com/news/2011-01-19/pickens-says-u-s-spent-337-billion-on-oil-imports-last-year.html.
7. National Renewable Energy Laboratory, "Dollars from Sense: The Economic Benefits of Renewable Energy," U.S. Department of Energy, September 1997, www.nrel.gov/docs/legosti/fy97/20505.pdf.

Chapter 3

1. Jill Longval, telephone interview by the author, July 26, 2011.
2. Bill Sheehan and Helen Spiegelman, "Climate Change, Peak Oil, and the End of Waste," in *The Post Carbon Reader: Managing the 21st Century's Sustainability Crises*, Richard Heinberg and Daniel Lerch, eds. (Healdsburg, CA: Watershed Media, 2010), pp. 374–75.
3. Tim Franks, "Cows Make Fuel for Biogas Train," BBC Newsnight, October 24, 2005, http://news.bbc.co.uk/2/hi/science/nature/4373440.stm.
4. For a full discussion of how our transportation system needs to transform, see Richard Gilbert and Anthony Perl, *Transport Revolutions: Moving People and Freight Without Oil*, second edition (Gabriola Island, BC: New Society Publishers, 2010).
5. The NEED Project, "Energy Consumption," Secondary Energy Infobook, 2001, p. 67, www.need.org/needpdf/infobook_activities/SecInfo/ConsS.pdf.

Chapter 4

1. Dive into a Carpool, Piedmont Authority for Regional Transportation, www.partnc.org/carpool.html.
2. Transportation Demand Management Testimonials, Piedmont Authority for Regional Transportation, www.partnc.org/TDMtestimonials.html.
3. Where Can I Find Car Sharing? CarSharing.net, www.carsharing.net/where.html.
4. Small Appliances, Consumer Energy Center, California Energy Commission, www.consumerenergycenter.org/home/appliances/small_appl.html.
5. Small Appliances, Green3DHome.com, www.green3dhome.com/YourHouse/Kitchen/SmallAppliances.aspx.
6. "How Much Energy Is Actually Used by the Clothes Dryer?" Project Laundry List, www.laundrylist.org/en/faq/35-general-laundry-questions/51--how-much-energy-is-actually-used-by-the-clothes-dryer.
7. Zolton Cohen, "How to Conserve Energy at Home," TLC, http://tlc.howstuffworks.com/home/how-to-conserve-energy-at-home4.htm.
8. Sarah Lozanova, "Going Solar in a Condo: How One Couple Overcame the Challenges to Install a Photovoltaic System at Their Condominium," *Solar Today*, American Solar Energy Society, July–August 2008, www.solartoday-digital.org/solartoday/20080708/?pg=44#pg44.

Chapter 5

1. Richard Gilbert and Anthony Perl, *Transport Revolutions: Moving People and Freight Without Oil* (Gabriola Island, BC: New Society, 2010).
2. "Alderson Community Energy Plan," Alderson, West Virginia, www.downstreamstrategies.com/alderson/alderson_commenergyplan_pamphle.pdf.
3. About PACE, Renewable Funding, https://www.renewfund.com/pace/pace-overview.

4. Closed-loop biomass uses dedicated energy crops, whereas open-loop biomass relies on waste sources such as residues from sawmills, agriculture, forests, urban manufacturing, and construction.

5. Renewable Electricity Production Tax Credit (PTC), Federal Incentives/Policies for Renewables & Efficiency, Database of State Incentives for Renewables & Efficiency, http://dsireusa.org/incentives/incentive.cfm?Incentive_Code=US13F.

6. Federal Production Tax Credit, Windustry, www.windustry.org/federal-production-tax-credit.

7. Business Energy Investment Tax Credit (ITC), Federal Incentives/Policies for Renewables & Efficiency, Database of State Incentives for Renewables & Efficiency, www.dsireusa.org/incentives/incentive.cfm?Incentive_Code=US02F&re=1&ee=1.

8. Mark Bolinger, "Community Wind: Once Again Pushing the Envelope of Project Finance," Environmental Energy Technologies Division, Ernest Orlando Lawrence Berkeley National Laboratory, January 2011, p. 6, http://eetd.lbl.gov/EA/EMP/reports/lbnl-4193e.pdf.

9. Ibid., pp. 9–10.

10. Shawn E. Marshall, "Forming a National Community Choice Aggregation Network," Galvin Electricity Initiative, December 2010, www.galvinpower.org/sites/default/files/Community_Choice_Aggregation_Report_Final_1-4-11.pdf.

11. Christopher Williams, "The Low Profit Limited Liability Company (L3C): A New Model for Cleantech?" Energy Collective, September 13, 2010, http://theenergycollective.com/christopherwilliams/43313/low-profit-limited-liability-company-l3c-new-model-cleantech.

12. Frayne Olson, "Are Cooperative Businesses Outdated?," North Dakota State University, www.ag.ndsu.nodak.edu/qbcc/NDCCC/FrayneOlsoned.htm.

13. Vermont Group Net Metering Team, Michael Dworkin, et al., "Vermont Group Net Metering Information & Guidelines," December 2010, pp. 14–15.

14. PUD Fast Facts, "PUDs: Owned by the People They Serve," Washington Public Utility Districts Association, www.wpuda.org/pdf/FastFacts3.pdf.

15. Ontario Sustainable Energy Association, press release, "Ontario Takes Historic Step Towards Energy Future," March 21, 2006, www.wind-works.org/FeedLaws/Canada/OSEAHistoricStep.html.

16. John Lorinc, "Ontario Issues $8 Billion in Renewable Energy Contracts," blog, *New York Times*, April 9, 2010, http://green.blogs.nytimes.com/2010/04/09/ontario-issues-8-billion-in-clean-energy-contracts.

17. Paul Gipe, "Grading North American Feed-in Tariffs," World Future Council, May 2010, p. 14, www.wind-works.org/FeedLaws/USA/Grading%20N.Am.%20FITs%20Report.pdf.

18. Toby D. Couture, "Renewable Energy Feed-in Tariffs: An Analytical View," California Energy Commission FIT Workshop, Sacramento, CA, May 28, 2009, www.energy.ca.gov/2009_energypolicy/documents/2009-05-28_workshop/presentations/01_Couture_Feed-in_Tariff_Wkshop_May_28_09.pdf.

19. See www.ferc.gov/industries/hydropower/gen-info/licensing/small-low-impact.asp.

20. Information about the Snohomish County PUD is from the company's website (www.snopud.com) and from a telephone interview with Steve Klein on September 21, 2011.

Chapter 6

1. For additional information on SMUD Solar Shares, visit http://nwcommunityenergy.org/solar/solar-case-studies/navajo-nation.

2. Joy Hughes, telephone interview by the author, September 22, 2011.

3. Gary Nystedt, telephone interview by the author, September 19, 2011.

4. The outputs of renewable energy projects usually have certain "green rights" associated with them. These rights are typically referred to as environmental attributes or renewable energy credits (RECs), although they are also called green tags, green energy certificates, or tradable renewable certificates. Whatever the name, environmental attributes generally include the right to be viewed as the owner or holder of the legal and market rights associated with the green aspects of the project.

5. Most of the background information in this case study comes from City of Ellensburg, "Ellensburg WA's Community Solar Array," Solaripedia, www.solaripedia.com/13/358/4722/ellensburg_wa_solar_gary_nystedt.html.
6. Most of the background information in this case study comes from "Invest in a Green Future with the Burlington Cohousing Solar Project, Project Summary for Potential Investors" (project proposal, Community Energy Exchange, L³C, Burlington, VT, November 8, 2010).
7. Don Schramm, telephone interview by the author, September 28, 2011.
8. A ballasted rack system uses heavy concrete weights to hold the mounting rack and PV panels in place.

Chapter 7

1. Terry Meyer, telephone interview by the author, September 26, 2011.
2. Much of this case study is based on Ryan Stockwell, Morgan Winters, and Dan Thiede, "Lessons & Concepts for Advancing Community Wind," Minnesota Project, December 2009, www.mnproject.org/pdf/TMP_Advancing-Community-Wind_Dec09.pdf.
3. Jon Folkedahl, telephone interview by the author, October 3, 2011.
4. Suzanne Pude, telephone interview by the author, October 5, 2011.
5. "Project Background," Fox Islands Wind Project, www.foxislandswind.com/background.html.
6. "Largest New England Coastal Wind Project Begins Power Generation," Fox Islands Wind Project November 17, 2009, http://www.foxislandswind.com/pdf/FIWPR111609.pdf.
7. "Update on the Fox Islands Wind Project," Fox Islands Wind Project April 7, 2001, www.foxislandswind.com/pdf/April2011FIECWindProjectUpdateFINAL.pdf.
8. Chip Farrington, telephone interview by the author, October 3, 2011.

Chapter 8

1. FERC Federal Energy Regulatory Commission, press release, "FERC, Colorado Sign Agreement on Small Hydropower Development," August 25, 2010, www.ferc.gov/media/news-releases/2010/2010-3/08-25-10.asp.
2. Head pressure is water pressure created by the vertical distance that the water falls from its intake to a storage tank, pressure-reducing valve, or (in this case) hydroelectric turbine. Head can be expressed as a vertical distance (in feet or meters) or as pressure in pounds per square inch (psi).
3. Jeff White, telephone interview by the author, October 6, 2011.
4. Water hammer, also known as hydraulic shock, is a pressure surge or wave resulting when water in motion is forced to stop or change direction suddenly. This pressure wave can cause noise, vibration, or in extreme cases pipe collapse.
5. Named after James B. Francis of Lowell, Massachusetts, who developed an improved design in 1848. Francis turbines, which come in many different sizes, are the most common water turbines in use today.
6. "Hydro Plant," City of Spearfish, South Dakota, www.cityofspearfish.com/publicworks/hydroplant.htm.
7. Cheryl Johnson, telephone interview by the author, October 11, 2011.
8. Mark Brandenburger, telephone interview by the author, October 18, 2011.
9. Some of the background material for this case study was based on information from the City of Hamilton website at www.hamilton-city.org/index.aspx?page=1. The specific hydro project references are no longer available.
10. The Kaplan turbine, an evolution of the Francis turbine, is a propeller-type water turbine with adjustable blades first designed by Austrian professor Viktor Kaplan in 1913. Kaplan turbines are now widely used around the world in high-flow, low-head electricity production.

11. This information is included in the Federal Energy Regulatory Commission original license for the City of Hamilton, Project No. 12667-003, June 25, 2008, www.ferc.gov/industries /hydropower/gen-info/licensing/issued-licenses.asp.
12. Dave Greber, "Hamilton Breaks Ground on $500M Hydroelectric Plant," *Oxford (Ohio) Press*, June 30, 2010, www.oxfordpress.com/news/oxford-news/hamilton-breaks-ground-on-500m -hydroelectric-plant-789006.html.
13. Kent Carson, telephone interview by the author, October 18, 2011.

Chapter 9

1. Jim Bishop, telephone interview by the author, October 25, 2011.
2. "Harney County Hospital," Sustainable Northwest, November 1, 2011, www.sustainablenorthwest.org/resources/biomass/harney-county-hospital.
3. Geoff Battersby, telephone interview by the author, November 1, 2011.
4. Revelstoke Community Energy Corporation, brochure, p. 3, www.revelstokechamber.com /pdf/RCEC%20brochure_100110.pdf.
5. A wet ton is two thousand pounds of biomass "as is." This is the sum of the dry weight of the material plus its moisture content.
6. No. 6 fuel oil, also known as bunker oil, is a thick, black, tarlike liquid that is the "leftover" product of crude oil after the more valuable hydrocarbons have been removed. It needs to be heated in order to flow through pipes and burner nozzles.
7. Some of the background material for this case study was based on information from the Middlebury College website at http://blogs.middlebury.edu/biomass.
8. Tom Corbin, telephone interview by the author, October 27, 2011.

Chapter 10

1. Johanna Miller, Brian Shupe, and Stephanie Smith, *Energy Planning & Implementation Guidebook for Vermont Communities*, Vermont Natural Resources Council and Vermont League of Towns and Cities, April 2011, p. 33, www.nh.gov/oep/programs/energy/documents /energy_planning_guidebook.pdf.
2. USDA, press release No. 0079.11, "On-Farm Renewable Energy Production Shows Tremendous Growth," February 23, 2011, www.usda.gov/wps/portal/usda/usdahome ?contentidonly=true&contentid=2011/02/0079.xml.
3. Lisa Gibson, "Watts from Wastewater," *Biomass Power & Thermal*, September 20, 2011, http://biomassmagazine.com/articles/5817/watts-from-wastewater.
4. Bill Sheehan and Helen Spiegelman, "Climate Change, Peak Oil, and the End of Waste," in *The Post Carbon Reader: Managing the 21st Century's Sustainability Crises*, Richard Heinberg and Daniel Lerch, eds. (Healdsburg, CA: Watershed Media, 2010), pp. 374–75.
5. Farm Power Northwest, home page, www.farmpower.com/index.html.
6. Kevin Maas, telephone interview by the author, October 20, 2011.
7. "Farm Power," Climate Trust, www.climatetrust.org/farm_power.html.
8. Diane Greer, "Codigestion and Cogeneration in Des Moines," *BioCycle* 52, no. 2, (2011), p. 38, www.jgpress.com/archives/_free/002278.html.
9. Ibid.
10. Mike DiMaggio, telephone interview by the author, November 11, 2011.
11. Mark Torresani, telephone interview by the author, November 11, 2011.

Chapter 11

1. Tom Doggett and Charles Abbott, "U.S. Ethanol Production to Rise as EPA Sets Share," Reuters, November 29, 2010, www.reuters.com/article/2010/11/29/us-usa-ethanol-epa-idUSTRE6AS61L20101129.
2. www.biofuels.coop/about.
3. Lyle Estill, telephone interview by the author, March 19, 2006.
4. Lyle Estill, telephone interview by the author, October 24, 2011.
5. See www.bq-9000.org.
6. The credit did, in fact, expire in 2012.
7. Lynn Benander, telephone interview by the author, November 21, 2011.
8. www.cooppower.coop/index.php/co-op-power-news.
9. Delayne Johnson, telephone interview by the author, November 17, 2011.
10. Kris Bevill, "Iowa Ethanol Plant Gets Grant for Bolt-on Cellulosic Facility," *Ethanol Producer Magazine*, June 22, 2011, www.ethanolproducer.com/articles/7908/iowa-ethanol-plant-gets-grant-for-bolt-on-cellulosic-facility.

Chapter 12

1. Bruce D. Green and R. Gerald Nix, "Geothermal—The Energy Under Our Feet: Geothermal Resource Estimates for the United States," National Renewable Energy Laboratory, Technical Report NREL/TP-840-40665, November 2006, www1.eere.energy.gov/geothermal/pdfs/40665.pdf.
2. The term *thermal watt* refers to thermal rather than electrical power produced.
3. Hildigunnur H. Thorsteinsson and Jefferson W. Tester, "Barriers and Enablers to Geothermal District Heating System Development in the United States," *Energy Policy* 38 (2010), pp. 803–13, www.ourenergypolicy.org/wp-content/uploads/2011/11/2008_12_Elsevier_EnergyPolicy_Hildigunnur_Tester_BarriersEnablersGeopthermalDistrictHeatingSystemDevelopmentUS.pdf.
4. The NEED Project, "Energy Consumption," Secondary Energy Infobook, 2001, p. 66, www.need.org/needpdf/infobook_activities/SecInfo/ConsS.pdf.
5. An artesian well produces water that flows upward under pressure.
6. Kevin Rafferty, "A Century of Service: The Boise Warm Springs Water District System," *GHC Bulletin*, August 1992, pp. 1–3, http://geoheat.oit.edu/pdf/bulletin/bi030.pdf.
7. Kent Johnson, telephone interview by the author, December 6, 2011.
8. Kenneth W. Neely, "Production History for the State of Idaho Capital Mall Geothermal System 1983–1994," *GHC Bulletin*, February 1996, pp. 19–22, http://geoheat.oit.edu/bulletin/bull17-1/art2.pdf.
9. R. Gordon Bloomquist, "The Veterans Administration Hospital District Heating System, Boise, ID," *GHC Bulletin*, December 2006, pp. 10–12, http://geoheat.oit.edu/bulletin/bull27-4/art4.pdf.
10. City of Boise, press release, "Traffic Advisory: Boise State Geothermal Project Underway," September 7, 2011, www.cityofboise.org/Departments/Public_Works/NewsReleases/2011/page66895.aspx.
11. For additional information on the Klamath Falls district heating system, see Greg Pahl, *The Citizen-Powered Energy Handbook* (White River Junction, VT: Chelsea Green Publishing, 2007).
12. Tonya "Toni" Boyd and John W. Lund, "Trials and Tribulations of the Oregon Institute of Technology Small-Scale Power Plant," *GHC Bulletin*, August 2011, pp. 6–9, http://geoheat.oit.edu/bulletin/bull30-2/art2.pdf.
13. Tonya "Toni" Boyd, telephone interview by the author, November 21, 2011.
14. Ann Felton, telephone interview by the author, December 22, 2011.
15. Chris Brawley Morgan, "Charity Builds Green in New Development," *Oklahoman*, March 8, 2008, www.noble.org/press_release/2008/inthenews/charity.html.
16. Aaron McRee, telephone interview by the author, December 8, 2011.
17. The higher the EER number, the more efficient the equipment.

18. The most common heat pump sizing measurement is the "ton," which has nothing to do with how heavy the pump is. This archaic term is a holdover from the days when refrigeration units were used mainly to produce ice for old-fashioned iceboxes that kept food cool. A "one-ton" unit could make a ton of ice in a day. In today's terms, a one-ton heat pump can generate twelve thousand Btu of cooling per hour at an outdoor temperature of ninety-five degrees Fahrenheit, or twelve-thousand-Btu heat output at forty-seven degrees.

Chapter 13

1. Paul Gipe, "Solar PV Leads Small Town into Solar Big Leagues," RenewableEnergyWorld.com, November 21, 2011, www.renewableenergyworld.com/rea/news/article/2011/11/solar-pv-leads-small-town-into-solar-big-leagues.
2. Mike Antheil, telephone interview by the author, December 9, 2011.
3. "Gainesville Renewable Energy Center (GREC)," American Renewables, www.amrenewables.com/our-projects/gainesville-renewable-energy-center.php.
4. The space occupied by the compacted waste and the cover material at a landfill is called a cell. There can be many daily layers within active landfill cells.
5. Angeljean Chiaramida, "Deal Inked to Build Big Solar Farm," NewburyportNews.com, November 5, 2011, www.newburyportnews.com/local/x627657400/Deal-inked-to-build-big-solar-farm/print.
6. "Mayor Steps Up to Take Carbon Challenge," Eco Flash, Greater Newburyport Eco Collaborative, April 1, 2010, www.ecocollaborative.org/newsletters/eco-flash-040110.
7. "Newburyport Clean Tech Center," Greater Newburyport Eco Collaborative, www.ecocollaborative.org/clean-tech-center.
8. "Resolute Marine Energy Signs an Agreement with Newburyport CleanTech Center," Resolute Marine Energy, April 8, 2011, www.resolutemarine.com/news/resolute-marine-energy-signs-an-agreement-with-newburyport-cleantech-center.
9. Miriam Zerbel, "Ein Dorf macht ernst," *Welt am Sonntag*, April 10, 2010, www.welt.de/print/wams/vermischtes/article13127993/Ein-Dorf-macht-ernst.html.
10. Wildpoldsried Village website, www.wildpoldsried.de/index.shtml?Energie.
11. Wildpoldsried Village website, www.wildpoldsried.de/index.shtml?wasserkraft.
12. Miriam Zerbel, "Ein Dorf macht ernst," *Welt am Sonntag*, April 10, 2010, www.welt.de/print/wams/vermischtes/article13127993/Ein-Dorf-macht-ernst.html.
13. Ibid.
14. Wildpoldsried Village website, www.wildpoldsried.de/index.shtml?biogas.
15. Ibid.
16. Wildpoldsried Village website, www.wildpoldsried.de/index.shtml?photovoltaik.
17. Wildpoldsried Village website, www.wildpoldsried.de/index.shtml?Holz.
18. Wildpoldsried Village website, www.wildpoldsried.de/index.shtml?pflanzenklaer.
19. Wildpoldsried Village website, www.wildpoldsried.de/index.shtml?thermo.
20. Susi Vogl, email interview by the author, December 16, 2011.
21. Miriam Zerbel, "Ein Dorf macht ernst," *Welt am Sonntag*, April 10, 2010, www.welt.de/print/wams/vermischtes/article13127993/Ein-Dorf-macht-ernst.html.

Chapter 14

1. John Farrell, "America's Energy Future a Battle Between Entrenched Utilities and Clean, Local Power," RenewableEnergyWorld.com, November 28, 2011, www.renewableenergyworld.com/rea/blog/post/2011/11/americas-energy-future-a-battle-between-entrenched-utilities-and-clean-local-power.
2. Mike New, telephone interview by the author, October 7, 2011.

3. Geoff Keith, Bruce Biewald, Ezra Hausman, Kenji Takahashi, Tommy Vitolo, Tyler Comings, and Patrick Knight, "Toward a Sustainable Future for the U.S. Power Sector: Beyond Business as Usual 2011," prepared for the Civil Society Institute, November 16, 2011, pp. 1–3, www.synapse-energy.com/Downloads/SynapseReport.2011-11.CSI.BBAU-2011.11-037.pdf.
4. Terry Meyer, telephone interview by the author, September 26, 2011.
5. Mike Antheil, telephone interview by the author, December 9, 2011.
6. Jon Folkedahl, telephone interview by the author, October 3, 2011.
7. Joy Hughes, telephone interview by the author, September 22, 2011.
8. Lynn Benander, telephone interview by the author, November 21, 2011.
9. Lyle Estill, telephone interview by the author, October 24, 2011.

Glossary

ABSORBER. In passive solar design, the generally hard, dark, external surface of the heat storage element.

ACTIVE SOLAR. Systems or strategies that make use of mechanical or other devices to harvest and use solar energy for the production of heat or electricity.

ALGAE. A simple rootless plant that grows in sunlit waters that can be used as a feedstock for biodiesel.

ANAEROBIC BACTERIA. Microorganisms that live and reproduce in an environment that does not contain oxygen.

ANEMOMETER. An instrument that measures the speed or force of the wind.

ARTESIAN FLOW. An artesian well flows naturally from an aquifer that is pressurized.

AXIAL FLOW TURBINE. A hydroelectric turbine design in which the shaft through the center of the turbine runs in the same direction as the water flow, much like a boat propeller.

BASELOAD. The minimum amount of electricity that a utility must make available to its customers (i.e., that must be constantly provided to the grid).

BATCH PROCESS. A method of making biodiesel or other fuels that relies on a specific, limited amount of inputs for a single batch.

BIOCNG. Compressed natural gas produced from waste organic matter.

BIODIESEL. A clean-burning fuel made from natural, renewable sources such as new or used vegetable oil or animal fats.

BIOFUEL. A liquid fuel such as ethanol, methanol, or biodiesel, made from biomass resources.

BIOGAS. Gaseous fuel, especially methane, produced by the bacterial degradation of organic matter.

BIOMASS. Plant material, including wood, vegetation, grains, or agricultural waste, used as a fuel or energy source.

BIOPOWER. Electricity generated by any number of biomass fuels.

BLACK START. The ability to help in restarting a power generating station after a major power failure.

BULB TURBINE. A hydroelectric turbine named after its bulbous shape that also contains the generator.

BTU. British thermal unit(s), a quantitative measure of heat equivalent to the amount of heat required to raise one pound of water by one degree Fahrenheit.

BUSINESS ENERGY INVESTMENT TAX CREDIT (ITC). A federal corporate tax credit that supports renewable energy projects.

CANOLA. The common term, especially in North America, for rapeseed that can be used as a feedstock for biodiesel.

CARBON DIOXIDE (CO_2). A product of combustion and a so-called greenhouse gas that traps the Earth's heat and contributes to global warming.

CARBON MONOXIDE (CO). A colorless, odorless, lethal gas that is the product of the incomplete combustion of fuels.

CATALYST. A substance that, without itself undergoing any permanent chemical change, facilitates or enables a reaction between other substances.

CELL. In a landfill, the space occupied by the compacted waste and the cover material. There can be many daily layers within active landfill cells.

CELLULOSE. A simple sugar and large component of the biomass of plants.

CELLULOSIC ETHANOL. A type of ethanol generally produced from the rigid cell wall material that makes up the majority of plants.

CLEAN RENEWABLE ENERGY BONDS (CREBs). A federal loan program for municipalities (including school districts), municipal utilities, and rural electric cooperatives to finance renewable energy projects.

CLOSED-LOOP SYSTEM. A geoexchange system using heat from the ground that is collected by means of a continuous loop of piping buried underground that is filled with antifreeze.

COMBINED HEAT AND POWER (CHP, ALSO KNOWN AS CO-GENERATION). A facility that generates electricity and thermal energy in a single, integrated system.

CONVENTIONAL OIL. Generally refers to free-flowing underground petroleum obtained naturally from its own pressure, physical lift, water flooding, and pressure from water or natural gas.

DIRECT USE. The use of geothermal hot water for direct heating of pools, greenhouses, fish farms, and other purposes.

DISTILLERS GRAIN. A by-product of the alcohol production process.

DISTRIBUTED GENERATION. An energy system in which smaller amounts of power are generated close to where it is used—typically a community, dwelling, or geographic region.

DISTRICT HEATING. A heating system for a group of homes or community that uses a central source of heat, normally distributed via underground pipes.

DIVERSION (OR RUN-OF-RIVER). A hydroelectric strategy that channels a portion of the river through a canal or pressurized pipe, which then spins a turbine connected to a generator, and eventually flows back into the river.

DRY CASK STORAGE. A temporary storage strategy for nuclear waste from nuclear power plants in the absence of a national storage facility.

E10 UNLEADED. Ordinary unleaded gasoline enhanced with ethanol, which is blended at a rate of 10 percent. E10 unleaded is approved for use by every major automaker in the world.

E85. A blend of 85 percent ethanol and 15 percent ordinary unleaded gasoline. This fuel mixture is used in flexible-vuel vehicles.

ELECTRICITY FEED-IN LAW (EFL). A law (or tariff) that requires electric utilities to make a connection with a renewable energy producer and to pay a specific long-term fixed price for the power produced.

ELECTROLYSIS. Using electricity to split molecules, such as water, into positive and negative ions (in the case of water, into hydrogen and oxygen ions).

EMISSIONS. All substances discharged into the air during combustion.

ENERGY BALANCE RATIO. A numerical figure that represents the energy stored in a fuel compared with the total energy required to produce, manufacture, transport, and distribute it.

ENERGY CROPS. Crops grown specifically for their energy value.

ENGINE/GENERATOR SET (GENSET). An internal combustion engine coupled with a generator used to produce electricity.

ENZYMATIC CATALYSIS. A new process, using enzymes, that can make use of low-quality fats and greases that are otherwise hard to transform into biodiesel.

EROEI (ENERGY RETURN ON ENERGY INVESTED). Ratio of the energy available in a produced resource to the energy that was used to produce it. Related to net energy.

ETHANOL. A colorless, flammable liquid that can be produced chemically from ethylene or biologically from the fermentation of various sugars from carbohydrates found in agricultural crops and residues from crops or wood. Also known as ethyl alcohol, alcohol, or grain spirits.

EXTREME ENERGY. The risks involved with extreme price volatility associated with declining supplies of, and increasing demand for, fossil fuels and their associated environmental costs.

FEED-IN TARIFF (FIT). A feed-in tariff is a policy mechanism designed to encourage investment in renewable energy technologies that generate electricity. This is accomplished by guaranteeing grid access and offering long-term contracts to renewable energy producers.

FEEDSTOCK. Any material converted to another form of fuel or an energy product.

FISCHER-TROPSCH. A process for making a liquid, diesel-like fuel from fossil fuels or biomass.

FLOW. The amount or volume of water that is available at a hydroelectric site.

FOSSIL FUEL. An organic, energy-rich substance formed from the long-buried remains of prehistoric organic life. These fuels are considered nonrenewable, and their use contributes to air pollution and global warming.

FRACKING. See hydraulic fracturing.

FRANCIS TURBINE. Named after James B. Francis of Lowell, Massachusetts, who developed an improved design in 1848. Francis turbines, which come in many different sizes, are the most common hydroelectric water turbines in use today.

GAS SCRUBBER. A system or device used to remove impurities from a gas.

GASEOUS EMISSIONS. Substances discharged into the air during combustion, typically including carbon dioxide, carbon monoxide, water vapor, and hydrocarbons.

GASIFICATION. A highly efficient process for converting biomass into gas (syngas, woodgas, and so on) through incomplete combustion in a specially designed firebox.

GASIFIER. A device designed to produce gas (syngas, woodgas, and so forth).

GEOTHERMAL. Earth heat or heat from the Earth, either high-temperature from deep within the Earth, or low-temperature from the top fifteen feet of the Earth's surface.

GEOTHERMAL EXCHANGE (GEOEXCHANGE). The movement of heat energy to and from the Earth to heat and cool an indoor environment.

GIGAWATT-HOUR (GWH). One gigawatt-hour equals one million kilowatt-hours.

GLOBAL WARMING. An increase in the average temperature of the Earth's atmosphere, especially a sustained increase sufficient to cause climate change.

GLYCERIN (OR GLYCEROL). A thick, sticky substance that is part of the chemical structure of vegetable oils and is a by-product of the process for making biodiesel. Refined glycerin is often used in the manufacture of soap and pharmaceuticals.

GREENHOUSE EFFECT. The heating of the atmosphere that results from the absorption of re-radiated solar radiation by certain gases, especially carbon dioxide and methane.

GRID INTER-TIED. An electrical system that is connected to the national electrical system or grid (see off-grid).

GROUP NET METERING. Allows renewable energy systems owned by a group of people to send excess power not immediately needed directly back into the electrical grid while crediting the participating producers for the excess power (see net metering).

HEAD. The water pressure created by the vertical distance that the water falls from its intake to the turbine in a hydroelectric installation.

HEAT EXCHANGER. Any device designed to transfer heat.

HEAT PUMP. A device that warms or cools a building by transferring heat from a relatively low-temperature reservoir to one at a higher temperature, aided by a compressor.

HIGH-TEMPERATURE GEOTHERMAL. Used mainly to generate electricity (geopower).

HOT WATER RESERVOIR. A geothermal resource that contains hot water.

HVAC. Heating, ventilation, and air-conditioning systems.

HYDRAULIC FRACTURING. The procedure of creating fractures in rock formations by injecting fluids into cracks to force them open to allow more gas and oil to flow out of the formation. Informally referred to as "fracking."

HYDROELECTRIC PLANT. An electricity production facility in which the turbine generators are driven by falling or flowing water.

HYDROPOWER. The use of moving water to provide mechanical or electrical energy.

INVERTER. An electrical device that converts direct current (DC) to alternating current (AC).

JEVONS PARADOX. Says that technological progress that increases the efficiency of the use of a resource tends to increase (rather than decrease) the rate of consumption of the resource.

KAPLAN TURBINE. A hydroelectric turbine design developed in 1913 by Viktor Kaplan featuring adjustable turbine blades that look like the propeller of a ship.

KILOWATT (KW). One thousand watts; ten 100-watt lightbulbs consume a kilowatt of electricity. One kilowatt equals 3,415 Btu.

KILOWATT-HOUR (KWH). A unit of energy equivalent to one kilowatt of power expended for one hour of time.

LANDFILL GAS (LFG). All of the gases generated by the decomposition of organic waste in a landfill in the absence of oxygen.

LARGE HYDROPOWER. A hydroelectric facility that generates more than thirty megawatts.

LIQUEFIED NATURAL GAS. Natural gas that has been condensed into a liquid by cooling to an extremely low temperature (-260 degrees Fahrenheit).

LIQUID BIOFUELS. Most commonly biodiesel, ethanol, and methanol, derived from biomass conversion.

LOCAL ENERGY. Local energy projects rely on locally available renewable energy resources that serve local needs.

LOSS ZONE. A stream zone, or segment, where flow is lost to underlying geological formations.

MEGAWATT (MW). One megawatt equals one million watts.

METHANE DIGESTER. A device, usually a large covered tank, containing organic matter that produces methane gas in the absence of oxygen (see anaerobic bacteria).

METHANOL. A volatile, colorless alcohol, originally derived from wood but also from fossil fuels, that is often used as a racing fuel and as a solvent. Also called methyl alcohol.

MICRO-HYDRO. A hydroelectric system that generates less than one hundred kilowatts.

MINNESOTA FLIP. A creative legal strategy in which the vast majority of ownership in a wind farm is initially placed in the hands of a large corporate entity that can benefit from the Federal Production Tax Credit during the first ten years. Then, after the credit has been fully utilized, the ownership reverts (or flips) back to the actual owners.

MODERATE TEMPERATURE GEOTHERMAL. Mainly used for district heating or other direct-use purposes.

NET ENERGY BALANCE. The difference between the energy produced and the energy it takes to produce it.

NET METERING (OR NET BILLING). Allows home-based renewable energy systems to send excess power not immediately needed in the home directly back into the electrical grid while crediting the homeowner for the excess power (see group net metering).

NET-ZERO ENERGY. Producing as much energy on an annual basis as is consumed on site, usually with renewable energy sources.

NEW MARKETS TAX CREDIT (NMTC). Originally intended to spur new or increased investments located in low-income communities, but has been used to help finance some renewable energy projects.

NITROGEN OXIDES (NOx). A product of combustion and a contributing factor in the formation of smog and ozone.

NO. 6 (OR BUNKER) FUEL OIL. A thick, black, tar-like liquid that is the "leftover" product of crude oil after the more valuable hydrocarbons have been removed, often used to fuel industrial or institutional heating systems.

NUCLEAR FISSION. The fission of uranium-235, an isotope of uranium, that supplies energy for nuclear reactors and atomic bombs.

OFF-GRID. An electrical system that is not connected to the national electrical system or grid (see grid inter-tied).

OIL SHALE. A black or dark brown shale containing hydrocarbons that yields petroleum when heated and processed.

OPEN SYSTEM. A geoexchange system that uses the latent heat in a body of water, which is usually a well but sometimes a pond or stream, as its heat source.

PARTICULATE EMISSIONS. Substances discharged into the air during combustion. Typically they are fine particles, such as carbonaceous soot and various organic molecules.

PASSIVE HOUSE. A passive house is a building in which a comfortable interior climate can be maintained without active heating and cooling systems. The house heats and cools itself, hence the term passive.

PASSIVE SOLAR. A broad category of solar techniques and strategies for regulating a building's indoor air and (sometimes) water temperatures.

PEAK COAL. The point at which global coal production reaches its all-time maximum and begins its terminal decline.

PEAK OIL. The point at which global oil production (especially production of conventional oil) reaches its all-time maximum and begins its terminal decline.

PELLET STOVE. A heating stove that is fueled with wood pellets.

PENSTOCK. A pipe that connects the water source and the turbine in a hydroelectric installation.

PHOTOSYNTHESIS. A process by which plants and other organisms use light to convert carbon dioxide and water into a simple sugar. Photosynthesis provides the basic energy source for almost all organisms.

PHOTOVOLTAIC (PV). Modules that utilize the photovoltaic effect to generate electricity.

POWER PURCHASE AGREEMENT (PPA). A contract that defines the selling prices for power and energy as well as the amount of power and energy sold, and includes provisions to ensure that performance does not fall below a certain standard.

PRODUCTION TAX CREDIT (PTC). The PTC reduces the federal income taxes of qualified tax-paying owners of renewable energy projects based on the electrical output (measured in kilowatt-hours).

PRODUCTS OF COMBUSTION. Substances formed during combustion. The products of complete fuel combustion are carbon dioxide and water. Products of incomplete combustion can include carbon monoxide, hydrocarbons, soot, tars, and other substances.

PROPERTY ASSESSED CLEAN ENERGY PROGRAMS (PACE). These programs enable local governments to finance renewable energy and energy-efficiency projects on residential, commercial, and industrial properties.

REACTION TURBINE. A hydroelectric turbine that operates fully immersed in water. Examples include Francis, propeller, and Kaplan.

REINJECTION (INJECTION) WELL. A geothermal reinjection well is any well constructed to dispose of geothermal fluids after the heat has been removed.

RELOCALIZATION. An increasingly popular grassroots strategy for rebuilding the local economy and infrastructure of communities so they may provide the basic necessities of life in an energy-constrained future.

RENEWABLE ENERGY. An energy source that renews itself or is renewable from ongoing natural processes.

RENEWABLE ENERGY PRODUCTION INCENTIVE (REPI). A federal performance-based incentive for projects owned by local municipal governments (including school districts), state governments, municipal utilities, rural electric cooperatives, and native corporations.

SECOND GENERATION BIOFUEL. Biofuels such as cellulosic ethanol and biomass Fischer-Tropsch diesel.

SHALE GAS. Natural gas found in shale deposits, generally produced via hydraulic fracturing (fracking). It has become an increasingly important, and controversial, source of natural gas in the United States.

SMALL HYDROPOWER. Hydroelectric installations that generate between one hundred kilowatts and thirty megawatts.

SMART GRID. An electricity transmission system (grid) that uses digital technology to improve reliability, resiliency, flexibility, and efficiency.

SOLAR COLLECTOR ARRAY. A group of solar collectors.

SOLAR GARDEN. A co-op, partnership, LLC, or other business entity, normally with multiple subscribers, that owns a community solar photovoltaic array.

SOLAR HOT WATER (SHW). An active water heating system that uses solar energy as the heat source to produce hot water for domestic purposes.

SUSTAINABLE. Used to describe material or energy sources that, if carefully managed, will provide at current levels indefinitely.

SWITCHGRASS. A summer perennial grass native to North America that is resistant to many pests and plant diseases and capable of producing high yields with very low applications of fertilizer, making it an ideal biomass feedstock.

SYNTHESIS GAS (ALSO KNOWN AS SYNGAS, PRODUCER GAS, BIOSYNGAS, GENGAS, OR WOODGAS). A synthetic gas produced through the gasification of biomass.

SYNTHETIC OIL. Oil manufactured using the Fischer-Tropsch process, which converts carbon dioxide, carbon monoxide, and methane into liquid hydrocarbons of various forms.

TAR SANDS. A mixture of tar (or bitumen), sand, and clay.

TECHNICAL GLYCEROL. A high-quality, valuable type of glycerol.

TIGHT SANDS GAS. Conventional natural gas extracted from unconventional reservoirs.

TIPPING POINT. In the context of peak oil, the tipping point is reached when global demand for oil finally exceeds the supply, causing exponential increases in the price for oil.

TON. The most common heat pump sizing measurement; a one-ton heat pump can generate twelve thousand Btu of cooling per hour at an outdoor temperature of ninety-five degrees Fahrenheit, or twelve-thousand-Btu heat output at forty-seven degrees.

TURBINE. A rotary engine, usually containing a series of vanes, blades, or buckets mounted on a central rotating shaft that is actuated by a current or stream of water, steam, or air.

UNCONVENTIONAL OIL. Bitumen, deepwater oil, extra-heavy oil, oil shale, shale oil, and tar sands (oil sands), all more difficult to extract and more expensive to refine compared with conventional oil.

USDA RURAL UTILITY SERVICE (RUS). Potentially makes low-cost debt financing available for a wide range of renewable energy projects, even those not associated with rural electric cooperatives.

WASTE VEGETABLE OIL (WVO). Used vegetable oil (normally from restaurants) that can be used as a feedstock for biodiesel, or burned directly as fuel in a retrofitted vehicle or heater.

WATER HAMMER. Also known as hydraulic shock, a pressure surge or wave resulting when water in motion is forced to stop or change direction suddenly. This pressure wave can cause noise, vibration, or in extreme cases pipe collapse in water systems.

WATER TURBINE. A rotary engine containing a series of blades or buckets mounted on a central rotating shaft that is actuated by a current or stream of water.

WAVE ENERGY. A free and sustainable energy resource created as wind blows over the ocean surface.

WAVE-POWERED ENERGY CONVERTER. A device that extracts energy directly from the surface motion of ocean waves and produces electricity.

WIND FARM. A group of wind turbines.

WIND TURBINE. A turbine for the generation of electricity that is powered by the wind.

WOOD PELLET. A fuel for heating normally made from highly compressed sawdust.

BIBLIOGRAPHY

Boxwell, Michael. *Solar Electricity Handbook, 2011 Edition: A Simple Practical Guide to Solar Energy —*
  Designing and Installing Photovoltaic Solar Electric Systems. Ryton on Dunsmore, Warwickshire,
  UK: Greenstream Publishing, fourth revised edition, 2011.
Boyle, Godfrey. *Renewable Energy*. New York: Oxford University Press, second edition, 2004.
Brown, Lester R. *Plan B 2.0: Rescuing a Planet Under Stress and a Civilization in Trouble*. New York:
  W. W. Norton & Company, 2006.
Chiras, Dan. *The Solar House: Passive Heating and Cooling*. White River Junction, VT: Chelsea Green
  Publishing, 2002.
Davis, Scott. *Microhydro: Clean Power from Water*. Gabriola Island, BC: New Society Publishers, 2004.
Deffeyes, Kenneth S. *Beyond Oil: The View from Hubbert's Peak*. New York: Hill and Wang, 2005.
Douthwaite, Richard. *Short Circuit: Strengthening Local Economies for Security in an Unstable World*.
  Totnes, UK: Green Books, 1996. (This classic is unfortunately out of print and hard to find.
  However, an online version containing the complete original text is available at www.feasta
  .org/documents/shortcircuit/contents.html.)
Doxon, Lynn Ellen. *The Alcohol Fuel Handbook*. West Conshohocken, PA: Infinity Publishing, 2001.
Estill, Lyle. *Industrial Evolution: Local Solutions for a Low Carbon Future*. Gabriola Island, BC: New
  Society Publishers, 2011.
———. *Biodiesel Power: The Passion, the People, and the Politics of the Next Renewable Fuel*. Gabriola
  Island, BC: New Society Publishers, 2005.
Gipe, Paul. *Wind Power: Renewable Energy for Home, Farm, and Business*. White River Junction, VT:
  Chelsea Green Publishing, 2004.
Harvey, Adam, Andy Brown, Priyantha Hettiarachi, and Allen Inversin. *Micro-Hydro Design
  Manual: A Guide to Small-Scale Water Power Schemes*. Rugby, UK: ITDG Publishing, 1993.
Heinberg, Richard. *The End of Growth: Adapting to Our New Economic Reality*. Gabriola Island, BC:
  New Society Publishers, 2011.
———. *Peak Everything: Waking Up to the Century of Declines*. Gabriola Island, BC: New Society
  Publishers, 2010.
———. *Blackout: Coal, Climate and the Last Energy Crisis*. Gabriola Island, BC: New Society Publishers, 2009.
———. *The Oil Depletion Protocol: A Plan to Avert Oil Wars, Terrorism and Economic Collapse*. Gabriola
  Island, BC: New Society Publishers, 2006.
———. *Powerdown: Options and Actions for a Post-Carbon World*. Gabriola Island, BC: New Society
  Publishers, 2004.
———. *The Party's Over: Oil, War and the Fate of Industrial Societies*. Gabriola Island, BC: New
  Society Publishers, 2003.
Heinberg, Richard, and Daniel Lerch, editors. *The Post Carbon Reader: Managing the 21st Century's
  Sustainability Crises*. Healdsburg, CA: Watershed Media, 2010.
Holmgren, David. *Permaculture: Principles and Pathways Beyond Sustainability*. Hepburn, Victoria,
  Australia: Holmgren Design Services, 2002.
Hopkins, Rob. *The Transition Companion: Making Your Community More Resilient in Uncertain Times*.
  White River Junction, VT: Chelsea Green Publishing, 2011.
Kachadorian, James. *The Passive Solar House*. White River Junction, VT: Chelsea Green Publishing,
  revised edition, 2006.
Komp, Richard J. *Practical Photovoltaics: Electricity from Solar Cells*. Ann Arbor, MI: Aatec Publica-
  tions, third edition, 2001.
Kunstler, James Howard. *The Long Emergency: Surviving the End of the Oil Age, Climate Change, and
  Other Converging Catastrophes of the Twenty-first Century*. New York: Atlantic Monthly Press, 2005.

Lerch, Daniel. *Post Carbon Cities: Planning for Energy and Climate Uncertainty.* Sebastopol, CA: Post Carbon Press, 2008.

Martenson, Chris. *The Crash Course: The Unsustainable Future of Our Economy, Energy, and Environment.* Hoboken, NJ: Wiley, 2011.

McKibben, Bill. *Eaarth: Making a Life on a Tough New Planet.* New York: St Martin's Griffin, 2011.

Pahl, Greg. *Biodiesel: Growing a New Energy Economy.* White River Junction, VT: Chelsea Green Publishing, second edition, 2008.

———. *The Citizen-Powered Energy Handbook: Community Solutions to a Global Crisis.* White River Junction, VT: Chelsea Green Publishing, 2007.

———. *Natural Home Heating: The Complete Guide to Renewable Energy Options.* White River Junction, VT: Chelsea Green Publishing, 2003.

———. *The Complete Idiot's Guide to Saving the Environment.* Indianapolis, IN: Alpha Books, 2001.

Potts, Michael. *The New Independent Home: People and Houses That Harvest the Sun.* White River Junction, VT: Chelsea Green Publishing, 1999.

Schumacher, E. F. *Small Is Beautiful: Economics as if People Mattered.* New York: Harper Perennial, reprint edition, 2010.

Simmons, Matthew. *Twilight in the Desert: The Coming Saudi Oil Shock and the World Economy.* Hoboken, NJ: Wiley, 2006.

Solar Energy International. *Photovoltaics: Design and Installation Manual.* Gabriola Island, BC: New Society Publishers, 2004.

Thomas, Dirk. *The Woodburner's Companion: Practical Ways of Heating with Wood.* Chambersburg, PA: Alan C. Hood & Co., second edition, 2004.

INDEX

Note: Page numbers in *italics* refer to figures and photographs.

About the Author

Greg Pahl has been involved in environmental issues for more than thirty years. He is a founding member of the Vermont Biofuels Association as well as the Addison County Relocalization Network (ACORN). A former military intelligence officer in the U.S. Army during the Vietnam War and a full-time freelance journalist for many years, he has written hundreds of articles and commentaries on energy and sustainable living. He is the author of five previous books, including *The Citizen-Powered Energy Handbook: Community Solutions to a Global Crisis* (2007).

About Post Carbon Institute

Post Carbon Institute provides individuals, communities, businesses, and governments with the resources needed to understand and respond to the interrelated economic, energy, environmental, and equity crises that define the twenty-first century.

About the Foreword Author

Van Jones is founding president of Rebuild the Dream, a pioneering initiative to restore good jobs and economic opportunity. He is the cofounder of three thriving organizations: the Ella Baker Center for Human Rights, Color of Change, and Green For All. Jones is also the author of two *New York Times* best sellers—*Rebuild the Dream* and *The Green Collar Economy*. A Yale-educated attorney, Jones worked as the green jobs advisor to the Obama White House in 2009. There, he helped run the interagency process that oversaw $80 billion in green recovery spending. Among numerous other honors, *Time Magazine* has named Jones one of the 100 most influential people in the world.

Get excerpts, resources, and more from the

Community Resilience Guides series at

resilience.org

 post carbon institute

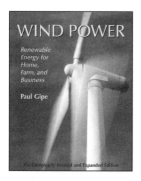